Molecular Biotechnology

Volume II

Molecular Biotechnology

Volume II

Ms Kusum Chopra

B.Tech (Biotechnology) (MIET) U.P. Tech.

Ms Hiru Ranabhatt

M.Sc. Plant Biotechnology from TERI University, New Delhi

WOODHEAD PUBLISHING INDIA PVT LTD

New Delhi

Published by Woodhead Publishing India Pvt. Ltd.
Woodhead Publishing India Pvt. Ltd.,
303, Vardaan House, 7/28, Ansari Road,
Daryaganj, New Delhi - 110002, India
www.woodheadpublishingindia.com

First published 2020, Woodhead Publishing India Pvt. Ltd.
© Woodhead Publishing India Pvt. Ltd., 2020

Woodhead Publishing India Pvt. Ltd. ISBN: 978-93-88320-27-6
Woodhead Publishing India Pvt. Ltd. e-ISBN: 978-93-88320-28-3

Typeset by Asian Enterprises, New Delhi
Printed and bound by Replika Press Pvt. Ltd.

Contents

Section III

Molecular biotechnology of eukaryotic system

Genetic engineering of plants

17.1   Introduction

Dramatic progress has been made in the development of gene transfer systems for higher plants. The ability to introduce foreign genes into plant cells and tissues and to regenerate viable, fertile plants has allowed for explosive expansion of our understanding of plant biology and has provided an unparalleled opportunity to modify and improve crop plants. Genetic engineering of plants offers significant potential for seed, agrichemical, food processing, specialty chemical and pharmaceutical industries to develop new products and manufacturing processes. The extent to which genetically engineered plants will have an impact on key industries will be determined both by continued technical progress and by issues such as regulatory approval, proprietary protection and public perception. The stable introduction of foreign genes into plants represents one of the most significant developments in a continuum of advances in agricultural technology that includes modern plant breeding, hybrid seed production, farm mechanisation and the use of agrichemicals to provide nutrients and control pests. The first-generation applications of genetic engineering to crop agriculture are targeted at issues that are currently being addressed by traditional breeding and agrichemical discovery efforts: (i) improved production efficiency, (ii) increased market focus and (iii) enhanced environmental conservation.

Genetic engineering methods complement plant breeding efforts by increasing the diversity of genes and germplasm available for incorporation into crops and by shortening the time required for the production of new varieties and hybrids. Genetic engineering of plants also offers exciting opportunities for the agri-chemical, food processing, specialty chemical and pharmaceutical industries to develop new products and manufacturing processes. The first transgenic plants expressing engineered foreign genes were tobacco plants produced by the use of *Agrobacterium tumefaciens* vectors.

Transformation was confirmed by the presence of foreign DNA sequences in both primary transformants and their progeny and by an antibiotic resistance phenotype conferred by a chimeric neomycin phosphotransferase gene. These early transformation experiments often utilised plant protoplasts as the recipient cells; the subsequent development of transformation methods based on regenerable explants such as leaves, stems and roots contributed significantly to the facile and routine transformation methods that are used today for many dicotyledonous plant species. A variety of free DNA delivery methods, including microinjection, electroporation and particle gun technology are being developed for the transformation of monocotyledonous plants such as corn, wheat and rice. In view of the rapid progress that is being made, it is likely that all major dicotyledonous and monocotyledonous crop species will be amenable to improvement by genetic engineering within the next few years.

17.2   Methods for introducing genes into plants

Agrobacterium tumefaciens-mediated gene transfer: Derivatives of the plant pathogen *Agrobacterium tumefaciens* have proved to be efficient, highly versatile vehicles for the introduction of genes into plants and plant cells. Most transgenic plants produced to date were created through the use of the *Agrobacterium* system. *Agrobacterium tumefaciens* is the etiological agent of crown gall disease and produces tumorous crown galls on infected species.

The utility of this bacterium as a gene transfer system was first recognised when it was demonstrated that the crown galls were actually produced as a result of the transfer and integration of genes from the bacterium into the genome of the plant cells. Virulent strains of *Agrobacterium* contain large Ti (for tumour inducing) plasmids, which are responsible for the DNA transfer and subsequent disease symptoms.

Genetic and molecular analyses showed that Ti plasmids contain two sets of sequences necessary for gene transfer to plants; one or more T-DNA (transferred DNA) regions that are transferred to the plant and the Vir (virulence) genes which are not, themselves, transferred during infection. The T-DNA regions are flanked by border sequences that were shown to be responsible for the definition of the region that is to be transferred to the infected plant cell. The T-DNA contains 8 to 13 genes, including a set for production of phytohormones, which are responsible for formation of the characteristic tumours when transferred to infected plants.

Early experiments demonstrated that heterologous DNA inserted into the T-DNA could be transferred to plants along with the existing T-DNA genes. Efficient plant transformation systems were constructed by removing the phytohormone biosynthetic genes from the T-DNA region, thereby eliminating the ability of the bacteria to induce aberrant cell proliferation. Modern plant transformation vectors are capable of replication in *Escherichia coli* as well as *Agrobacterium*, allowing for convenient manipulations.

The general features of these vectors and the process of transfer to plant cells are outlined in Fig. 17.1. Recent technological advances in vectors for *Agrobacterium* mediated gene transfer have involved improvements in the arrangements of genes and restriction sites in the plasmids that facilitate construction of new expression vectors. Vectors in current use have convenient

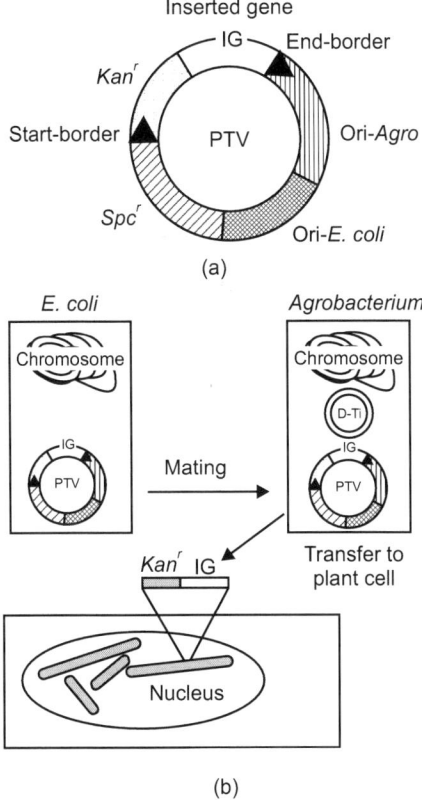

(a)

(b)

Figure 17.1: *Agrobacterium*-mediated plant transformation. (A) Generalised plant transformation vector (PTV). The plasmid contains an origin of replication that allows it to replicate in *Agrobacterium* (Ori-Agro) and a high copy number origin of replication functional in *E. coli* (Ori-*E. coli*). This allows for easy production and testing of engineered plasmids in *E. coli* prior to transfer to *Agrobacterium* for subsequent introduction into plants. Two resistance genes are usually carried on the plasmid, one for selection in bacteria, in this case for spectinomycin resistance (*Spc*^{r}) and the other that will express in plants, in this example encoding kanamycin resistance (*Kan*^{r}). Also present are sites for the addition of one or more inserted genes (IG) and directional T-DNA border sequences which, when recognised by the transferfinctions of *Agrobacterium*, delimit the region that will be transferred to the plant. (B) Diagram of the plant transformation process. The PTV constructed in *E. coli* is transferred to an engineered *Agrobacterium* by a 'triparental' mating procedure. The engineered *Agrobacterium* contains a 'disarmed' Ti plasmid (D-Ti) from which the genes necessary for pathogenesis have been removed. Virulence functions on the D-Ti interact in trans with the border sequences on the PTV mobilising the region between them into a plant cell and inserting it into one of the plant's chromosomes within the nucleus. The kanamycinresistant phenotype conferred by the *Kan*^{r} gene allows the selection of transformed plant cells during plant regeneration.

multilinker regions, which may be flanked by a promoter and a polyadenylate addition site for direct expression of inserted coding sequences.

Agrobacterium constitutes an excellent system for introducing genes into plant cells, since (i) DNA can be introduced into whole plant tissues, which by-passes the need for protoplasts and (ii) the integration of T-DNA is a relatively precise process. The region of DNA to be transferred is defined by the border sequences; occasional rearrangements do occur, but in most cases an intact T-DNA region is inserted into the plant genome. This contrasts with free DNA delivery systems in which the plasmids routinely undergo rearrangment and concatenation reactions before insertion and can lead to chromosomal rearrangements during insertion in both animal and plant systems. Sequencing of insertion sites shows that only small duplications or other changes occur in flanking sequences during T-DNA integration.

The stability of expression of most genes that are introduced by *Agrobacterium* appears to be excellent. Published studies have shown that integrated T-DNAs give consistent genetic maps and appropriate segregation ratios. Introduced traits have been found to be stable over at least five generations during cross-breeding and seed increase on genetically engineered tomato and oilseed rape plants. This stability is critical to the commercialisation of transgenic plants. The list of plant species that can be transformed by *Agrobacterium* has been greatly expanded and now includes several of the most important broadleaf crops (Table 17.1).

Table 17.1: Species for which the production of transgenic plants have been reported. Abbreviations: (At) *Agrobacterium tumefaciens,* (Ar) *Agrobacterium rhizogenes*, (FP) free DNA introduction into protoplasts, (PG) particle gun, (MI) microinjection, (IR) injection of reproductive organs.

Plant species	*Method*
Herbacious dicots	
Petunia	At
Tomato	At
Potato	At
Tobacco	At
Arabidopsis	At
Lettuce	At
Sunflower	At
Oilseed rape	At
Flax	At
Cotton	At
Sugarbeet	At

(Cont'd...)

Plant species	Method
Celery	At
Soyabean	At
Alfalfa	At
Medicago varia	At
Lotus	At
Vigna aconitifolia	FP
Cucumber	Ar
Carrot	Ar
Cauliflower	Ar
Horseradish	Ar
Morning glory	Ar
Woody dicots	
Poplar	At
Walnut	At
Apple	At
Monocots	
Asparagus	At
Rice	FP
Corn	FP
Orchard grass	FP
Rye	IR

Advances in other transformation technologies: In those systems where *Agrobacterium*-mediated transformation is efficient, it is the method of choice because of the facile and defined nature of the gene transfer. Few monocotyledonous plants appear to be natural hosts for *Agrobacterium*, although transgenic plants have been produced in asparagus with *Agrobacterium* vectors and transformed tumours have been observed in yam. Cereal grains such as rice, corn and wheat have not been successfully transformed by *Agrobacterium*, despite encouraging evidence for T-DNA transfer in corn.

Extensive efforts have consequently been directed toward the development of systems for the delivery of free DNA into these species. The first of these systems to give demonstrable transformation of plant cells relied on physical means similar to those used in the transformation of cultured animal cells. Transformation has been achieved in plant protoplasts through facilitation of DNA uptake by calcium phosphate precipitation, polyethylene glycol treatment, electroporation, or combinations of these treatments. These methods have allowed the production of transgenic cells for the study of gene expression in systems that cannot be transformed by other means.

The applicability of these systems to the production of transgenic plants is limited by the difficulties involved in regenerating plants from protoplasts. There have been significant advances in the regeneration of cereals (traditionally one of the most recalcitrant groups) from protoplasts. Several laboratories have succeeded in regenerating fertile rice plants from protoplasts. This advance was rapidly followed by the production of transgenic rice plants through the delivery of free DNA to protoplasts followed by regeneration. Progress in regeneration of corn has been more limited; one group demonstrated regeneration of mature plants from protoplasts and succeeded in producing transgenic plants. However, all plants were sterile, apparently as a result of the necessary period in culture or the regeneration procedure.

While this progress is encouraging, limitations remain in the application of this technology to cereal crop improvement. In corn and rice, the ability to form regenerable protoplasts appears to be primarily confined to a small number of varieties. Even if the fertility problems are overcome, introduction of the transferred genes into the broad range of commercial varieties in use today would require a lengthy period of backcrossing.

In parallel with the work on protoplast transformation, efforts to find novel ways to introduce DNA into intact cells or tissues have been emphasised. Regeneration of cereals from immature embryos or from explants is relatively routine. One of the most significant developments in this area has been the introduction of 'particle gun' or high-velocity microprojectile technology. In this system, DNA is carried through the cell wall and into the cytoplasm on the surface of small (0.5 to 5 μm) metal particles that have been accelerated to speeds of one to several hundred meters per second.

The particles are capable of penetrating through several layers of cells and allow the transformation of cells within tissue explants. Production of transformed corn cells and fertile, stably transformed tobacco and soyabean plants with particle guns has already been demonstrated.

By eliminating the need for passage through a protoplast stage, the particle gun method has the potential to allow direct transformation of commercial genotypes of cereal plants. Intensive efforts to produce transgenic cereals by the use of particle guns are currently under way in many laboratories around the world.

Other methods that have the potential to influence the production of transgenic cereals include gene transfer into pollen, direct injection into reproductive organs, microinjection into cells of immature embryos and rehydration of desiccated embryos. There has been some demonstration of transient or stable gene expression through the use of each of these methods in some species, but the range of their applicability remains to be demonstrated.

17.3   Application of genetic engineering to crop improvement

The availability of efficient transformation systems for crop species is of intense interest to biotechnology, agrichemical and seed companies for the application of this technology to crop improvement. Initial research has been focused on the engineering of traits that relate directly to the traditional roles of industry in farming, such as the control of insects, weeds and plant diseases. Progress has been rapid and genes conferring these traits have already been successfully introduced into several important crop species.

Weed control: Engineering herbicide tolerance into crops represents a new alternative for conferring selectivity and enhancing crop safety of herbicides. Research has largely concentrated on those herbicides with properties such as high unit activity, low toxicity, low soil mobility and rapid biodegradation and with broad spectrum activity against various weeds. The development of crop plants that are tolerant to such herbicides would provide more effective, less costly and more environmentally attractive weed control. The commercial strategy in engineering herbicide tolerance is to gain market share through a shift in herbicide use–not to increase the overall use of herbicides, as is popularly held. Herbicide-resistant plants will have the positive impact of reducing overall herbicide use through substitution of more effective and environmentally acceptable products.

Two general approaches have been taken in engineering herbicide tolerance: (i) altering the level and sensitivity of the target enzyme for the herbicide and (ii) incorporating a gene that will detoxify the herbicide. As an example of the first approach, glyphosate, the active ingredient of Roundup herbicide, acts by specifically inhibiting the enzyme 5-enolpyruvylshikimate-3-phosphate synthase (EPSPS). Glyphosate is active against annual and perennial broadleaf and grassy weeds, has very low animal toxicity and is rapidly inactivated and degraded in all soils. Tolerance to glyphosate has been engineered into various crops by introducing genetic constructions for the overproduction of EPSPS or of glyphosate-tolerant variant EPSPS enzymes.

Similarly, resistance to sulphonylurea compounds, the active ingredients in Glean and Oust herbicides, has been produced by the introduction of mutant acetolactate synthase (ALS) genes. Glean and Oust are broad-spectrum herbicides and are effective at low application rates. Since both EPSPS and ALS activities are present in wild-type plants, the possibility of deleterious effects on crop performance or product quality due to their reintroduction is unlikely. The use of these herbicides in new crop applications may require re-examination of residues of the herbicides; however, since the residue safety levels for these two compounds in food crops have already been established, this is not an issue unique to genetically engineered plants. Resistance to gluphosinate and bromoxynil has been achieved by the alternative approach of introducing bacterial genes encoding enzymes that inactivate the herbicides by acetylation or nitryl hydrolysis, respectively. In field tests the gluphosinate-tolerant plants have shown excellent tolerance to the herbicide. Evaluation of the biological activity of the specific herbicide conjugates and metabolites that may be present in the transgenic plants will be carried out according to existing chemical residue regulations.

Current crop targets for engineered herbicide tolerance include soyabean, cotton, corn, oilseed rape and sugarbeet. Factors such as herbicide performance, crop and chemical registration costs, potential for out-crossing to weed species, proprietary rights issues and competing herbicide technologies must all be considered before final decisions on commercialisation of specific herbicide-tolerant crops can be made.

Insect resistance: The production of insect-resistant plants is another application of genetic engineering with important implications for crop improvement and for both the seed and agrichemical industries. Progress in engineering insect resistance in transgenic plants has been achieved through the use of the insect control protein genes of *Bacillus thuringiensis* (*Bt*). *Bacillus thuringiensis* is an entomocidal bacterium that produces an insect control protein which is lethal to selected insect pests. Most strains of *Bt* are toxic to lepidopteran (moth and butterfly) larvae, although some strains with toxicity to coleopteran (beetle) or dipteran (fly) larvae been described. The insect toxicity of *Bt* resides in a large protein; this protein has no toxicity to beneficial insects, other animals, or humans. The mode of action of the *Bt* insect control protein is thought to be exerted at the level of disruption of ion transport across brush border membranes of susceptible insects.

Transgenic tomato, tobacco and cotton plants containing the *Bt* gene exhibited tolerance to caterpillar pests in laboratory tests. The level of insect control observed in the field tests with tobacco and tomato plants has been excellent; in one such test tomato plants containing the *Bt* gene suffered no agronomic damage under conditions that led to total defoliation of control plants.

The excellent insect control observed under field conditions indicates that this technology may have commercial application in the near future. Early market opportunities for caterpillar resistance are leafy vegetable crops, cotton and corn. Crop targets for beetle resistance are potato and cotton. Other types of insecticidal molecules are necessary to extend biotechnology approaches for controlling additional insect pests in these and other target crops. Plants genetically engineered to express a proteinase inhibitor gene are partially resistant to tobacco budworm in laboratory experiments, field tests will be necessary to determine the agronomic utility of this approach.

Disease resistance: Significant resistance to tobacco mosaic virus (TMV) infection, termed 'coat protein-mediated protection,' has been achieved by expressing only the coat protein gene of TMV in transgenic plants. This approach produced similar results in transgenic tomato, tobacco and potato plants against a broad spectrum of plant viruses, including alfalfa mosaic virus, cucumber mosaic virus, potato virus X and potato virus Y. One mechanism of coat protein-mediated cross protection appears to involve interference with the uncoating of virus particles in cells before translation and replication.

Transgenic tomatoes carrying the TMV coat protein gene have been evaluated in greenhouse and field tests and shown to be highly resistant to viral infection. The transgenic plants showed no yield loss after virus inoculation, whereas the yield was reduced 23% to 69% in control plants. The level of capsid protein in the engineered plants [typically 0.01% to 0.5% of the total protein] is well below the levels found in plants infected with this endemic virus. This fact should facilitate registration and commercialisation of virus-resistant plants. Virus resistance could provide significant yield protection in important crops such as vegetables, corn, wheat, rice and soyabean.

While limited success in engineering resistance to fungal diseases has been reported, genetically engineered resistance to fungal pathogens and to bacteria remains in the early research stages.

17.4 Key advances in expression and gene isolation technology

Dramatic progress has been made in our understanding of and ability to alter the regulation of gene expression in plants and in techniques for the identification and isolation of genes of interest. In many cases, this progress has been facilitated by the availability of efficient gene transfer systems. The engineered plants discussed in the previous section generally depend on the use of continuously expressed promoters driving dominant single gene traits. Future plant genetic engineering will probably include alteration of traits that require subtle temporal and spatial regulation of gene expression and introduction or alteration of entire biosynthetic pathways.

Regulated gene expression: Genes that show precise temporal and spatial regulation in leaves, floral organs, seeds and other plant organs have now been identified and isolated from a number of species of higher plants. Within the next few years, genetic engineers will have in hand a large battery of regulatory sequences that will allow for accurate targeting of gene expression to specific tissues within transgenic plants. In addition, a number of genes that respond to external influences, such as heat shock, anaerobiosis, wounding, nutrients and applied phytohormones, have been isolated and characterised. The control regions of these genes may also find utility in genetic engineering strategies. The ability to decrease the expression of a gene in a transgenic plant also has potential utility in the study of plant gene expression and function as well as in crop improvement. Significant successes have already been achieved with genes that produce antisense RNAs to the messengers for polygalacturonase in tomato fruits and chalcone synthase in petunia and tobacco plants. In all of these studies, substantial reductions (up to 90%) in the levels of the mRNA and protein products of the target genes were observed. Striking phenotypic alterations were observed in some of these transgenic plants.

This method of constructing mutant phenotypes will significantly enhance biochemical and physiological studies on protein and enzyme function. In an alternative approach to reducing expression of a gene, the enzymatic regions derived from self-splicing RNA molecules are used to design RNA enzymes capable of specific RNA cleavage. *In vitro* studies have demonstrated the potential of this method, but it has yet to be applied in plants. Preliminary work on insertion of donor DNA into plant chromosomes by homologous recombination indicates that it may also be possible to use this approach for the selective inactivation of a gene.

Gene tagging: Advances in methods for the identification and isolation of new gene coding sequences are of great importance to the engineering of improved plants. The cloning of transposon sequences has allowed the isolation of genes from several species by transposon-mediated gene tagging. The demonstration that mobile elements isolated from maize are able to transpose when introduced into dicot species indicates that this powerful technique is applicable to any plant species for which transformation is possible. It has also been shown that under appropriate transformation conditions, the T-DNA of a plant transformation vector can itself serve as an insertional mutagen.

Gene mapping: Major efforts have been mounted to obtain high resolution restriction fragment length polymorphism (RFLP) genetic maps in a number of plant species. The availability of such a map in tomato has already led to the resolution of several loci affecting quantitative quality traits. The RFLP mapping technique will be especially powerful in Arabidopsis where the small genome size and lack of significant repetitive sequences will simplify the process of genome 'walking' from an RFLP marker to a closely linked gene. The availability of Arabidopsis genomic libraries in cosmids, which can also act as plant transformation plasmids, will allow direct testing of the isolated DNA for its ability to complement the mutation of interest at each step of the walking process. In addition, such libraries may be used in large-scale transformation experiments to directly rescue genes by complementing mutants with a selectable phenotype.

17.5　Key issues affecting introduction of genetically engineered plants

The advances in crop improvement by genetic engineering have occurred so rapidly that the initial introduction of these crops in the marketplace will be primarily influenced by nontechnical issues. These issues include regulatory approval, proprietary protection and public perception.

Regulatory approval: In the United States, genetically engineered plants potentially come under the statutory jurisdiction of three federal agencies: The United States Departmlent of Agriculture (USDA), Food and Drug Administration (FDA) and Environmental Protection Agency (EPA). The field testing of genetically engineered crops has been less controversial than the introduction of other recombinant organisms into the environment. In the last few years there have been over a dozen tests of engineered crops in diverse locations across the United States-by year end there will be over 30 such tests. All of these tests have been reviewed in detail by the USDA, with input from the other government agencies The key consideration in approval of these tests has been a scientific evaluation of the risk and environmental impact of a particular field test experiment. Several studies and discussions of the issues and perceptions that surround the release of genetically engineered crops have produced a consensus that such engineered crops present virtually no direct risk to human or animal health. The specific knowledge of the introduced DNA sequences, the detailed understanding of the known functions of the gene products and the high level of biological or physical containment were cited as key reasons for the inherent low risk to human and animal health.

The 'success' of such small field tests, while important, has over shadowed other needs in the regulatory process. For example, many unanswered questions remain regarding the cost and regulatory requirement for large-scale utilised field tests. It is important that an approval process be developed to accommodate the rapid transition that will occur as testing of engineered crops goes from small, isolated field plots to large-scale, utilised testing; the development of genetically engineered crop varieties and hybrids will ultimately occur in the fields around the world-not in the research laboratory. The mechanism for FDA or EPA approval or endorsement of genetically engineered plants and food products remains undefined. Issues such as regulatory requirements, registration costs and commercialisation are already becoming significant issues for companies attempting to develop improved genetically engineered crops fortemid-1990s. Several groups, such as the International Food Biotechnology Council (IFBC) and the Federation of American Scientists for Experimental Biology (FASEB) expert panel on criteria for determining the regulatory status of food and food ingredients produced by new technologies, consisting of academic scientists and representatives of major food, chemical, biotechnology and seed companies, are working with government agencies to develop appropriate registration guidelines. The regulation of transgenic plants must be based on scientific principles that: (i) meet the general publics need for a safe and reasonably priced food supply and (ii) recognise the inherent low risk of gene transfer technology and the benefits afforded by genetically engineered crops to growers, food processors and consumers.

Proprietary protection: Patent protection for genetically engineered plants is considered essential to offset the cost of developing crops with significant new traits. The Supreme Court decision in Diamond v. Chakrabarty ruled that micro-organisms were not unpatentable simply because they were living cells and in 1985, the U.S. Board of Patent Appeals and Interferences ruled specifically that whole plants were patentable. Numerous companies have since filed patent applications that cover the genes, the processes of isolating genes and making the genetically modified plants and seeds themselves. Patent protection provides a broader proprietary right than is provided under either the International Union for the Protection of New Varieties of Plants (UPOV) or the U.S. Plant Variety Protection Act (PVPA). The scope of the proprietary right of a patent on a plant is broadened by the absence of the experimental use exceptions found in protection afforded by plant varietal certification status. Although no one disputes that companies that have invested heavily in R&D to isolate, test and commercialise genes are entitled to protection for their inventions, there is considerable debate with in the seed industry concerning how much protection is deserved and what impact patents will have on the cooperative nature of the seed industry itself. The concern has been voiced that patents on plants will favour large seed companies and reduce the overall number of companies. In contrast, while there were three private soyabean seed companies before PVPA, now there are more than, patenting plants will likely create finrther incentive to invest in the seed industry in order to position it to meet the technological challenges and supply needs of the future. Much of this debate results from confusion surrounding the restrictions imposed by patent rights versus the incentive they provide for the competitive research and product development that stimulates innovation. Many of the conciliatory proposals, including patenting of genes (but not plants) and compulsory licensing in the event that plant patenting is permitted, if implemented, could significantly reduce the incentive for private industry funding in this field.

Lack of proprietaryprotection for genetically engineered plants outside the United States remains a serious limitation; plant and animal varieties are largely excluded from patent protection by European countries that signed the 1973 European Patent Convention. At this time only specific processes can be patented. The European Patent Office (EPO) is currently readdressing the patenting of plants and animals, but this seems certain to be appealed and it may be several years before the situations clear and only then will begin the wave of oppositions, appeals and infringement actions that have marked the early pharmaceutical patents in the biotechnology area. Enforceability of plant patents in other countries, including Japan, China and Eastern Bloc countries, is questionable. While there are numerous initiatives to harmonise both registration and proprietary protection throughout the key trading countries in the world, the outcome is not imminent and will be unlikely to have an impact on first generation products.

Public perception: Genetically engineered crops are being developed at a time when a lack of understanding regarding the importance of agricultural research exists. Current issues, including concerns about (i) periodic, temporary production surpluses, (ii) changing farm infrastructure, (iii) inconsistency in farm policies and (iv) a general distrust for new technologies, have at times overshadowed the long-term need for the provision of economical, high-quality food products for a growing world population. Currently, at the beginning of the 1989 cropping season, world reserves of grain are at their lowest level since the years immediately following World War II; another drought in 1989 could create a world food emergency.

Despite this background, recent polls conducted by the Office of Technology Assessment indicate that most people believe that the benefits of agricultural biotechnology research outweigh remote risks. In view of the initial public debate that has occurred over the last several years on field testing and environmental release of genetically engineered organisms, it would seem that agricultural biotechnology has indeed passed its first major public perception obstacle.

The next test of the public acceptance of this technology will come in several years when food products derived from genetically engineered crops enter the general food supply. The current focus on issues of risk and environmental release has heightened the need for increased science education and open discussion of issues. It is essential that the safety and benefits of agricultural biotechnology research and the critical role that it will play in providing for world food demand be communicated and understood, so that informed decisions by the public are possible.

17.6 Future prospects on genetically engineered plants

During the last 10 years, the availability of gene transfer systems has catalysed a major refocusing on plants as a biological system, the use of genetically engineered plants as an analytical tool to explore unique aspects of gene regulation and development and the potential to produce novel commercial crop varieties has created a high level of scientific excitement and has driven research into many new areas. The breadth of information to be gained from the study of transgenic plants is serving as an important focus for unifying basic plant science research in plant breeding, pathology, biochemistry and physiology with molecular biology. Regulation of gene expression is the fundamental basis for manipulating cellular metabolism and this new research tool offers the possibility of extending physiological and genetic observations to a mechanistic level. In the next few years we can expect to see major advances in our understanding of basic plant processes.

These advances, in turn, will accelerate the application of genetically engineered plants in the seed production and agrichemical industries. The major crops that can currently be improved with genetic techniques are soyabean, cotton, rice and alfalfa (Table 17.1). The timing of commercialisation of genetically engineered crops is ultimately determined by the need to address each of the following issues: (i) evaluation of field performance, (ii) breeding and seed increase for commercial-scale release, (iii) establishment of optimal agronomic practices and (iv) regulatory approval and crop certification.

The worldwide agrichemical industry has been and will continue to be a leading sponsor of agricultural biotechnology research. All major agrichemical companies have R&D efforts in the area of biotechnology for crop improvement. These companies see opportunities to develop new products and extend the use of existing products, as well as to be positioned at the leading edge of new technologies that may have a significant impact on existing agrichemical businesses.

Genetic engineering of plants also offers exciting opportunities for the food processing industry to develop new products and more cost-effective processes. While many of the early successful examples of genetically engineered plants have focused on agronomic genes, it is possible that the food processing and specialty chemical industries may represent the greatest commercial opportunity for biotechnology. Examples of such applications include production of: (i) larger quantities of starch or specialised starches with various degrees of branching and chain length to improve texture and storage properties, (ii) higher quantities of specific oils or the elimination of particular fatty acids in seed crops and (iii) proteins with nutritionally balanced amino acid composition. The ability to reduce processing costs by the elimination of anti-nutritive or off-flavour components in foods is quite feasible with antisense nucleic acid technology. The enzymes and genes involved in biosynthesis of colouring materials and flavours are important to the food industry and to the consumer. Studies on the biosynthesis of some of these compounds have been hampered by the low quantities of enzymes present in the producing cells, but new techniques based on gene tagging may overcome these difficulties.

Enormous opportunity lies in the successful use of crops for both commodity and specialty chemical products. Plants have traditionally been a source of a wide range of polymeric materials. These range from starch and celluloses, which are carbohydrate-based, to polyhydrocarbons such as rubber and waxes. Many of these polymers have been replaced in the last two to three decades by synthetic materials derived from petroleum-based products. However, the cost, supply and waste-stream problems often associated with petroleum-based products are issues that are focusing renewed attention on the use of biological polymers. Genetic engineering will significantly enlarge the spectrum and composition of available plant polymers.

Plants also offer the potential for production of foreign proteins with various applications to health care. Proteins such as neuropeptides, blood factors and growth hormones could be produced in plant seeds and this may ultimately prove to be an economical means of production. Several mammalian proteins have been produced in genetically engineered plants and expression of pharmaceutical peptides in oilseed rape plants has been reported.

Transgenic plants

18.1 Introduction

Transgenic plants, or plants which express foreign gene products, can be generated by a variety of procedures, such as *Agrobacterium*-mediated transformation or biolistic delivery. Recent advances in these technologies have resulted in the development of commercially successful disease and herbicide resistant plants which both increase crop yield and reduce costs for the farmer. Safe and inexpensive production of recombinant proteins in large quantities have also been produced in transgenic plants, as well as plants which possess enhanced nutritional traits. In this chapter, current techniques employed in plant transformation are investigated. Transgenic plants resistant to plant viruses, insects and herbicides are discussed.

There is virtually no place on earth where the term 'transgenic plant', referring to plants that contain foreign genetic material, is unfamiliar. Transgenic plants which exhibit increased pest and disease resistance can prevent global production losses which are currently greater than 35%. Transgenic plants also present enormous possibilities to become one of the most cost-effective and safe systems for the large-scale production of proteins for industrial, pharmaceutical, veterinary and agricultural uses. In these cases, the plant derived protein must be biologically identical to its native counterpart and be produced at levels high enough to be purified by relatively simple procedures.

18.2 Transformation of plants

Plant transformation, meaning the stable integration of the gene of interest into the plant genome, was originally conducted using a modified strain of *Agrobacterium tumefaciens*, the bacterial strain responsible for crown-gall disease. *Agrobacterium tumefaciens* harbours a large tumour inducing (Ti) plasmid and during infection causes a mass of mainly undifferentiated cells to form on a plants stem at the soil line (crown).

The transfer DNA (T-DNA) portion of the Ti plasmid and its delimiting right and left border sequences become integrated into the nuclear genome of a susceptible plant cell that is in contact with the bacterium. The T-DNA encodes enzymes for synthesising plant hormones that stimulate cell division and the proliferation of undifferentiated cells into the tumour. Vectors used for transformation today lack the genes for hormone-synthesising enzymes and therefore can introduce foreign DNA into a nuclear chromosome of a plant cell with minimal damage.

Insertion of DNA into a plant by *A. tumefaciens* involves insertion of a foreign gene between the borders of the T-DNA, which in turn is cloned within a small plasmid. The construct is then transformed into a modified version of *A. tumefaciens* which lacks the virulence genes. Upon infection, the T-DNA is transferred into the plant cell and the gene of interest is incorporated into the host chromosome. The plant cell can then be regenerated from tissue culture into a mature transgenic plant by transferal through a series of culture media with different hormone contents.

A number of problems exist with this mode of transformation. Primarily, the restricted host range of *Agrobacterium* renders infection of monocots difficult. For this reason, other transformation procedures have been developed. Maize, for example, is commonly transformed by particle bombardment, a procedure in which high velocity microprojectiles carrying DNA can be 'shot' with compressed gas using a 'gene gun' into plant tissue.

In addition to this, foreign gene expression in nuclear transformed plants can vary markedly from one transgenic plant to another. Chromosomal position effects are partially responsible for this problem, since the insertion of the transgene into the plant genome is uncontrolled. Other difficulties include the ability of nuclear transformed plants to express more than one transgene. Since many agronomic traits are in fact multigenic and stem from the action of several genes, the production of transformants expressing multiple genes is a painstakingly long process.

More recently, genes have been introduced directly into the plastid genome. This was first accomplished for *Chlamydomonas reinhardtii* by biolistic transformation. Plastid transformation is unique from nuclear transformation as the transgene is incorporated directly into the plastid genome by homologous recombination and can be predictably directed to a specific site within the plastid chromosome. Recently, two new procedures involving polyethylene glycol and direct *in situ* injection have also been developed for plastid transformation.

Since chloroplast genes are arranged on operons, chloroplast transformation can be used to produce multicistronic mRNAs and in the future, traits determined by multiple genes can be expressed in chloroplasts. Transgene expression levels can be several fold higher in chloroplast-transformed plants than in their nuclear-transformed counterparts and lack the same variation in expression levels. The sequestration of foreign proteins in chloroplasts prevents their adverse interactions with the cytoplasmic environment and protects the cell from the accumulation of potentially toxic proteins. Since chloroplasts are not present in pollen, transgenes cannot be transferred to nearby sexually compatible crops to produce 'superweeds'. The ability of chloroplast transformation to overcome several major problems associated with conventional nuclear technologies has created unprecedented opportunities for plant biotechnology in the future.

18.3   Insect resistant plants

Tools of molecular biology and genetic engineering have provided humankind with unprecedented power to manipulate and develop novel crop genotypes towards a safe and sustainable agriculture in the 21st century. Technologies and chemical inputs that have proven harmful to human health and environment need to be replaced with safer alternatives to manage insect pests in agricultural ecosystems. Many insecticidal proteins and molecules are available in nature which are effective against agriculturally important pests but are innocuous to mammals, beneficial insects and other organisms. Insecticidal proteins present in *Bacillus thuringiensis (Bt)*, which have shown efficacy as spray formulations in agriculture over the past five decades, have been expressed in many crop species with positive results. Large scale cultivation of *Bt*-crops raises concerns about the possible development of resistant insects. Many strategies have been formulated to prevent/delay the development of resistance.

These strategies have to be given serious consideration where the first *Bt*-crop containing resistance to insect pests, particularly *Helicoverpa armigera*, has been released for commercial cultivation in the farmers fields. In addition to *Bt*, proteinase inhibitors present in several plant species offer a good source of resistance to insect pests. A combination of proteinase inhibitors has been suggested as a viable alternative to *Bt* to manage insects such as *H. armigera*. In recent years, several novel insecticidal proteins have been discovered in bacteria such as *Photorhabdus luminescens*. The judicious expression of multiple insecticidal proteins that differ in their mechanisms of toxicity will provide formidable barriers for insects to develop resistance. Finally, deployment of integrated pest management (IPM) strategies during the cultivation of transgenic crops will ensure durable insect resistance.

Insect pest menace is the major factor that destabilises crop productivity in agricultural ecosystems. A variety of insect pests ranging from lepidopterans to orthopterans damage crops and stored seed. The rich biodiversity of agricultural, horticultural and forest species faces a perpetual onslaught by insect infestation because of the predominantly tropical and sub-tropical climates prevalent in various countries. A survey conducted among plant breeders, pathologists and entomologists shows that breeding for resistance to insect pests is at the top of the priority list of many important crops. Table 18.1 lists some of the important pests on major crops of India. Improvement of crop productivity by the introduction of high-yielding varieties which are more responsive to applied nitrogen and lack of proper crop rotation practices has also resulted in an enhancement of pest incidence. Insect pest management by chemicals obviously has brought about considerable protection to crop yields over the past five decades.

Table 18.1: Important pests of major crops.

Crop	*Insect pest*		*Family*
Rice	Yellow stem borer	*Scirpophaga incertulas*	Lepidoptera
	Brown plant hopper	*Nilaparvata lugens*	Hemiptera
Mustard	Mustard aphid	*Lipaphys erysimi*	Hemiptera
Chickpea	Gram pod borer	*Helicoverpa armigera*	Lepidoptera
Pigeonpea	Gram pod borer	*H. armigera*	Lepidoptera
Cotton	Cotton boll worm	*H. armigera*	Lepidoptera
Sugarcane	Top borer	*S. nivella*	Lepidoptera
Groundnut	Leaf miner	*Stomopterix nertaria*	Lepidoptera
Potato	Tuber moth	*Phthorimaea operculella*	Lepidoptera
Tomato	Fruit borer	*H. armigera*	Lepidoptera
Brinjal	Shoot and fruit borer	*Leucinodes orbonalis*	Lepidoptera
Cauliflower and cabbage	Diamondback moth	*Plutella xylostella*	Lepidoptera

Unfortunately, extensive and very often, indiscriminate usage of chemical pesticides has resulted in environmental degradation, adverse effects on human health and other organisms, eradication of beneficial insects and development of pest-resistant insects. As we enter the new millennium with the objective of achieving higher and stable crop productivity to feed the burgeoning population, it is imperative to apply safe and environment friendly strategies to attain our goals. Insect pest management in an eco-friendly manner is no longer a dream. A large number of insecticidal molecules which are effective against insects and innocuous to man and other organisms have been reported. Tools of molecular biology and genetic engineering can facilitate harnessing and deployment of these molecules in crop plants in a safe and sustainable fashion. In this chapter the efficacy of various categories of insecticidal proteins for the development of insect-resistant transgenic plants are discussed.

18.4   Insecticidal proteins of *Bacillus thuringiensis*

Bacillus thuringiensis (*Bt*) is a Gram-positive, aerobic, sporulating bacterium which synthesises crystalline proteins during sporulation. These crystalline proteins are highly insecticidal at very low concentrations. As these proteins are non-toxic to mammals and other organisms, *Bt* strains and their insecticidal crystal proteins (ICPs) have acquired acceptability as eco-friendly biopesticides all over the world and have been under extensive use in agriculture, horticulture, forestry, animal health and mosquito control for the past four decades. With the advent of molecular biology and genetic engineering, it has become possible to use *Bt* more effectively and rationally by introducing the ICPs of *Bt* in crop plants. *Bt* strains and ICPs were first found to affect a range of lepidopteran insects, which are recognised worldwide as major agricultural pests on crops. Subsequently, discovery of new strains expanded the host range. Strains are now available which are toxic to coleopterans, dipterans, lice, mites and even nematodes. Most families of Lepidoptera include species susceptible to the *cry*1 and *cry*2 crystal proteins produced, in particular, by *Bt* serotypes *kurstaki* and *aizawai*. Currently, the crystal toxins are classified on the basis of amino acid sequence homology. The ICPs fall under 40 different classes with some toxins exhibiting specificity to multiple insect orders. Toxicity of various ICPs towards different pests has been studied and catalogued. Extensive screening programmes are in progress as *Bt* ICPs have high commercial value. The mechanism of action of the *Bt* ICPs has been worked out in some detail.

The molecular structure of at least three different ICPs has been studied. The crystals, upon ingestion by the insect larva, are solubilised in the highly alkaline midgut into individual protoxins which vary from 133 to 138 kDa in molecular weight, depending upon the type of protoxin. The protoxins are acted upon by midgut proteases which cleave them into two halves, the *N*-terminal half which is usually of 65–68 kDa is the toxin protein. The toxin protein fragment can be divided into three domains. The first is involved in pore formation, the second determines receptor binding and the third is involved in protection to the toxin from proteases. The toxin protein binds to specific receptors present in the midgut epithelial membranes. Upon receptor binding, the domain inserts itself into the membrane leading to the pore formation. The disturbances in osmotic equilibrium and cell lysis lead to insect paralysis and death.

The delivery of *Bt* ICPs through spray formulations, engineered *Bt* and other bacteria has certain limitations. The biopesticidal sprays suffer from short half-life, physical removal (wind and rain) and inability to reach burrowing insects. Engineered bacteria very often proliferate at a rate and quantity not sufficient to kill the target insects. These disadvantages can be overcome if the ICPs are expressed in the plant cells at levels sufficient enough to kill the larvae. The tobacco plants engineered with truncated genes encoding *cry*1Aa and *cry*1Ab toxins were found to be resistant to the larvae of tobacco hornworm. However, the levels of *cry* protein expression in the plant tissues were not very high. A significant breakthrough was made in 1990 by researchers at Monsanto Company (U.S.) who modified the *cry* genes (*cry1Ab and cry1Ac*) for better expression in plant cells. The codon usage of prokaryotic genes of *Bt* was altered to resemble that of higher plants. In addition, many features like presence of putative polyA type signals and splice sites which destabilise *Bt* mRNAs in plant cells were removed without altering the amino acid sequence of the ICPs. Expression of such modified genes in crop plants, *cry1Ac* in cotton and *cry3Aa* in potato, conferred considerable protection against lepidopteran and coleopteran pests respectively. Subsequently, many crop plants which include rice, maize, peanut, soyabean, canola, tomato and cabbage were transformed with various modified *cry* genes. An interesting example of native gene (*cry1IA5*) expression resulting in significant resistance to *H. armigera* in transgenic tobacco was provided by Selvapandiyan and others.

Another important landmark is the introduction of a native *cry1Ac* gene into the choloroplast genome of tobacco which expressed the *cry* protein to a very high level (3–5% of leaf soluble protein). Chloroplast transformation besides providing high foreign protein expression also ensures maternal transmission of the foreign gene and therefore avoiding the spread of transgene through pollen. If extended to the important crop plants such as cotton and rice, this strategy can prove very useful in future.

However, it remains to be seen if transformed chloroplast genomes will provide protection in the reproductive parts and fruiting bodies which are often the targets of insect attack. As of now, more than 30 plant species have been transformed with *Bt cry* genes (Table 18.2). The commercialisation of *Bt*-crops started in 1996 with the introduction of bollworm-resistant cotton ('Bollgard') in U.S. Subsequently, potato and maize were also commercialised. In India, intensive efforts are underway to introduce *cry* genes in crop plants such as rice, potato, cotton, sorghum and vegetables.

Table 18.2: Some important *Bt*-transgenic crops.

Crop	Gene	Target pests
Cotton	*cry1Ab/cry1Ac*	Bollworms
Corn	*cry1Ab*	European corn borer
Potato	*cry3a*	Colorado potato beetle
Rice	*cry1Ab/cry1Ac*	Stem borers and leaf folders
Tomato	*cry1Ac*	Fruit borers
Potato	*cry1Ab*	Tuber moth
Eggplant	*cry1Ab/cry1B*	Shoot and fruit borer
Canola	*cry1Ac*	Diamondback moth
Soyabean	*cry1Ac*	Soyabean looper
Corn	*cry1H/cry9C*	European corn borer

Note: Other crop species carrying various *cry* genes include peanut, alfalfa, cranberry, rutabaga, apple, white clover, white spruce, broccoli, grapevine, walnut, pear and sugarcane.

Investigations concerning evaluation of different ICPs for their relative toxicity to various target pests were made. Transgenic crop species carrying different *cry* genes are at various stages of development.

The first transgenic plants of tobacco developed at the Tata Energy Research Institute by using modified *cry1Ab* and *cry1C* showed considerable protection against tobacco caterpillar (*Spodoptera litura*) in limited field trials conducted at the Central Tobacco Research Institute. Scientists at the Bose Institute (Kolkata) have introduced a modified *cry1Ac* gene in rice (IR 64) for resistance to yellow stem borer. However, field evaluation of these rice transgenics has not been undertaken. A synthetic *cry1Ac* gene was introduced in rice (Pusa Basmati 1, Karnal Local and IR 64) under the control of Ubiquitin promoter and transgenic lines exhibiting total protection against neonate larvae of yellow stem borer (YSB) were identified. Field evaluation of these transgenics was performed in 2002 and lines resistant to YSB were identified. Vegetable crops such as brinjal and tomato were transformed by synthetic/modified *cry1Ab* and *cry1Ac* genes, respectively, to confer resistance to fruit borers.

Limited field trials of *Bt*-brinjal and *Bt*-tomato were conducted for three and two growing seasons respectively. The degree of insect protection was 75% and 94% in brinjal and tomato respectively. Four genotypes of potato were transformed by modified *cry1Ab* to achieve considerable protection against tuber moth and *H. armigera*. In addition to the work described above, many public and private sector institutions are involved in the development of insect resistant rice, cotton, sorghum, groundnut, sunflower, castor and tobacco. In the private sector, MAHYCO in collaboration with Monsanto introduced the modified *cry1Ac* gene originally used to transform Coker 312 variety of cotton into parental lines of hybrids that have been bred specially for Indian agronomic conditions. These transfers required four back-crosses and two selfed generations. The hybrids were field evaluated at different locations. Various experiments related to gene flow, effects of pollen and plants on non-target organisms, etc., were conducted. The results showed that *Bt*-cotton required no or minimal pesticide sprays while the non-transgenic plants required nine to twelve sprays to manage bollworms.

Commercial release of *Bt*-cotton has been approved by the Government of India in March 2002. Another seed company Nunhems-ProAgro Seeds has been conducting field trials of *Bt*-vegetables such as tomato and cauliflower, which carry modified *cry* genes and the results are awaited. There is a need for systematic evaluation of the insecticidal efficacy of *Bt* ICPs to pests such as *S. litura*, *Earias insulana*, *Chilo partellus*, *Spilosoma obliqua* (Fig. 18.1), *Maruca testulalis*, etc., as has been done in the case of *H. armigera* and *L. orbonalis*. Biochemical analysis of receptor binding *vis-à-vis* δ-endotoxins

Figure 18.1: *Spilosoma obliqua.*

could provide valuable information that can help design suitable toxin combinations to be expressed in transgenic plants.

18.4.1 Vegetative insecticidal proteins of *Bt*

Research efforts in the past five years have led to the discovery of novel insecticidal proteins which are produced by certain isolates of *B. thuringiensis*. These proteins unlike well-characterised crystal proteins are produced during vegetative growth of cells and are secreted into the growth medium. These proteins have been termed as vegetative insecticidal proteins (*vip*). Sequences encoding for a *vip* have been cloned, sequenced and the protein has been expressed in *E. coli*. The 88 kDa vegetative insecticidal protein has a putative bacillar secretory signal at the *N*-terminal which is not processed during its secretion. It does not show any homology with the known crystalline insecticidal proteins. This structural dissimilarity is indicative of a possible divergent insecticidal mechanism than the other known *Bt*-toxins. In experiments wherein the expressed receptor to *Bt*-toxin of polyphagous pest *S. litura* was titrated against *vip* toxin no interaction between these ligands was observed. These preliminary results together with the observed structural divergence of *vip* with other toxins make them an ideal candidate for deployment in insect management programmes together with the other category of *Bt*-toxins described earlier. Individually *vip* has been successfully expressed in monocots and dicot plants and efforts to pyramid *vip* in the *Bt*-transgenic crops are under way in several laboratories.

Other insecticidal proteins from bacteria, plants and animals

Proteinase inhibitors: Plants have a wide array of defense proteins including the proteinaceous proteinase inhibitors and lectins induced in response to insect attack. Proteinase inhibitors (PIs) represent the most well studied class of plant defense proteins and are abundantly present in the storage organs (seeds and tubers). Their role against herbivory was hypothesised due to their abundance and their lak of activity against endogenous proteases. However, in many cases, the transgenically expressed PIs have not demonstrated any resistance against insects (Table 18.3).

Table 18.3: Examples of transgenic plants expressing genes encoding proteinase inhibitors, α-amylase inhibitors and lectins.

Crop	Gene	Target pest
Tobacco	Cowpea serine PI	Tobacco bud worm
Tobacco	Potato serine PI	Tobacco hornworm
Rice	Cowpea serine PI	Stemborer
Potato	Cowpea serine PI	Lacanobia
Potato	Oryzacystatin	Potato beetle
Tobacco	Hornworm PI	Whitefly
Pea	Bean a-AI	Bruchids
Potato	Snowdrop lectin	Potato aphid
Rice	Snowdrop lectin	Brown plant hopper

Plant lectins

Lectins are found in most types of beans, including soyabeans. Reduced growth, diarrhea and interference with nutrient absorption are caused by this class of toxicants. Different lectins have different levels of toxicity, though not all lectins are toxic, though no all are toxins. Lectins interaction with certain carbohydrate is very specific. This interaction is as specific as the enzyme-substrate, or antigen-antibody interactions. Lectins play an important role in the defence mechanisms of plants against the attack of micro-organisms, pests and insects. Fungal infection or wounding of the plant seems to increase lectins. In legumes, the role of lectins in the recognition of nitrogen-fixing bacteria *Rhizobium genus*, which have sugar containing substances, has received a special attention.

Other functions of lectins in plants may include:

1. Enzymes (but unknown substrate).
2. Storage of proteins.
3. Defense mechanism.
4. Cell wall extension.
5. Mitogenic stimulation.
6. Transport of carbohydrates.
7. Packaging and/or mobilisation of storage materials.

Insect chitinases

Chitin, an insoluble structural polysaccharide that occurs in the exoskeletal and gut linings of insects, is a metabolic target of selective pest control agents. One potential biopesticide is the insect molting enzyme, chitinase, which degrades chitin to low molecular weight, soluble and insoluble oligosaccharides.

Insecticidal viruses

There are many viruses pathogenic to insect pests. These viruses are used in insect pest management programmes. Genomes of small viruses can be introduced into crop plants, which will synthesise the viral particles and acquire entomocidal property. For instance, *H. armigera* Stunt Virus (HaSV) is a tetravirus specific to lepidopteran insects and is very remotely related to viruses of plants and animals.

Genes from bacteria other than Bt

Another bacterium which aroused interest in recent years is *Photorhabdus luminescens* that dwells inside the gut of entomophagous nematodes, which belong to the family Heterorhabtidae.

Novel genes of plant origin

Cloning of genes from higher plants resistant to insect pests is feasible by a molecular breeding approach. Recent example of cloning of the *Mi-1* gene from wild tomato (*Lycopersicon peruvianum*) has given an opportunity to control root-knot nematode and potato peach aphid simultaneously. The vast biodiversity of Indian flora can yield rich dividends in this respect.

18.4.2　Perspectives

The management of insect pests in agriculture is feasible in a safe and effective manner. Molecular tools give us an opportunity to develop genotypes that carry resistance traits. The resistance needs to be protected by taking lessons from our past experiences with chemical pesticides. *Bt* has rightly emerged as a powerful tool of plant protection in agriculture in a sustainable manner. Although not universal in its application and total in its protection, *Bt* will play a central role in protecting the crop from its major insect pests. In combination with other powerful biopesticidal proteins such as proteinase inhibitors, *Bt* will drastically reduce the consumption of chemical pesticides and thus protecting the environment. Proteinase inhibitors and lectins have a major role to play in the management of secondary pests which are not susceptible to *Bt* and also as part of gene pyramiding strategies. It would also be appropriate if a particular *Bt* gene highly specific to a target insect is not deployed in multiple crops.

Basic plant molecular biology research is necessary to identify effective promoters which can sustain foreign protein expression during the late reproductive phase of crop plants such as cotton boll development. The durability of insect resistance in transgenic crops can only be ensured if integrated pest management (IPM) practices are followed. *Bt* as a biopesticidal formulation will continue to play an important role as a component of IPM in crop species which are not amenable to the attempts of genetic transformation.

18.5   Fungus resistant plants

This section deals with fungal pathogens of crop plants. Starting from the first step of mutual recognition of host and pathogen which involves resistance gene–avirulence gene interaction, moving onto immediate response of the plant in terms of hypersensitive response, production of active oxygen species, followed by local resistance response in terms of production of pathogenesis-related proteins and other antifungal proteins, then to the final step of systemic acquired resistance (SAR), all this information has been/or is being used to produce fungus resistant transgenic plants in different crop species. This chapter also discusses strategies that have been used to produce fungus resistant transgenic plants and also discuss some of the emerging possibilities in the wake of large scale genome sequencing aspects being undertaken in crop plants. Significant yield losses due to fungal attacks occur in most of the agricultural and horticultural species. Fungal diseases are rated either the most important or second most important factor contributing to yield losses in our major cereal, pulse and oilseed crops. On the basis of a recent survey, contribution of fungal diseases towards total yield loss in some important crops in India has been summarised in Table 18.4. Incidence of plant diseases has been controlled by agronomic practices that include crop rotation and use of agrochemicals and by breeding new strains and varieties that contain new resistance conferring genes.

Table 18.4: Contribution of fungal diseases.

Crop	Pathogen	Disease
Rice	*Pyricularia ozyzae*	Blast
Wheat	*Puccinai recondiata*	Leaf rust (Brown rust)
Maize	*Helminthosporium maydis* and *H. turcicum*	Leaf blight
Sorghum	*Sphacelotheca reiliaria*	Grain mould
Pigeonpea	*Fusarium udum*	Wilt
Chickpea	*Fusarium oxysporum*	Wilt
Brassica	*Alternaria brassiceae*	Blight
Soyabean	*Phakospora packyrhizi*	Rust
Potato	*Phytophthora infestans*	Late blight

The use of agrochemicals poses many dangers that include harmful effects on the ecosystem and an increase in the input cost of the farmers. The breeding of resistant crops is time consuming and has to be a continuous process as often new races of pathogens evolve and crops become susceptible. Despite the boom and bust cycles, breeders have been successful in protecting some of the major crops grown around the world from fungal diseases. A major success story is wheat in which systematic breeding has been done to develop varieties resistant to wheat rust by first incorporating genes from the primary gene pool and when this option ran out, from the secondary and tertiary gene pools of alien species and genera. Although shown to be possible, wide hybridisation programmes face numerous difficulties. Often sexual crosses are difficult to make and genetic exchange in the hybrids is poor due to low frequency of pairing between chromosomes of crop species and alien species. Problems can also arise due to linkage drag (genes for resistance are linked to some deleterious genes which lower the yield of the crop variety). Table 18.5 shows Pathogenesis related (PR) protein genes used for making fungus-resistant transgenic plants.

Novel alternative strategies that would circumvent the problems faced in wide hybridisation are required to produce fungus resistant crop varieties. Such strategies will be particularly important in cases where source of resistance is not available in taxonomically related species. The most significant development in the area of varietal development for disease resistance is the use of the techniques of gene isolation and genetic transformation to develop transgenics resistant to fungal diseases. Improvements in genetic transformation technology have allowed the genetic modification of almost all important food crops like rice, wheat, maize, mustard, pulses and fruits. To identify the important genes which need to be introduced in the plants to improve their resistance to fungal pathogens, lot of basic work has been done in the area of host – pathogen recognition. During the last decade, many resistance genes whose products are involved in recognising invading pathogens have been identified and cloned. A number of signalling pathways which follow the pathogen infection have been dissected. Table 18.6 shows two genes used in combination for making fungus-resistant transgenic plants.

Many of the antifungal compounds which are synthesised by plants to combat fungal infections have been identified. The complete sequencing of *Arabidopsis* genome has led to identification of a number of tentative resistance gene clusters. All this knowledge would greatly advance development of different strategies for producing fungus resistant transgenic plants. The production of fungus resistant transgenics can be basically classified into two categories namely: (i) production of transgenic plants with antifungal molecules like proteins and toxins and (ii) generation of a hypersensitive response through

Table 18.5: PR protein genes used for making fungus-resistant transgenic plants.

Plant species	PR protein	Donor	Fungus tested	Resistance
Alfalfa (Medicago sativa)	PR2 (class II glucanase)	Alfalfa (M. sativa)	Phytophthora megasperma	+
Canola (Brassica napus)	PR3 (class I chitinase)	Bean (Phaseolus vulgaris)	Rhizoctonia solani	+
			Pythium aphanidermatum	
Carrot (Daucus carota)	PR5	Tobacco (Nicotiana tabacum)	Erysiphe heraclei	+
Grapevine (Vitis vinifera)	PR3 (class I chitinase)	Rice (Oryza sativa)	Elisinoe ampelina	+
Kiwifruit (Actinidia chinensis)	PR2 (class I glucanase)	Soyabean (Glycine max)	Botrytis cinerea	+
Potato (Solanum tuberosum)	PR5	Potato (S. commersonii)	Phytophthora infestans	+
	PR5	Tobacco	P. infestans	+
Rapeseed (B. napus)	PR3 (class I chitinase)	Tobacco–tomato (chimeric)	Cylindrosporium concentricum	+
			Phoma lingam	+
			Sclerotinia sclerotiorum	+
Rice	PR3 (class I chitinase)	Rice	Rhizoctonia solani	+
		Rice	Magnaporthe grisea	+
	PR5	Rice	Rhizoctonia solani	+
Tobacco (N. tabacum)	PR1a	Tobacco	Pernospora tabacina	+
			Phytophthora parasitica var. nicotianae	+
			Cercospora nicotianae	−

Plant species	PR protein	Donor	Fungus tested	Resistance
	PR2 (class I glucanase)	Soyabean	*P. infestans*	+
	PR2 (class II glucanase)	Alfalfa	*C. nicotianae*	+
		Barley (*Hordeum vulgare*)	*R. solani*	+
	PR3 (class I) chitinase	Bean	*R. solani*	+
		Rice	*C. nicotianae*	+
		Tobacco	*R. solani*	+
	PR3 (class II chitinase)	Barley	*R. solani*	+
	PR3 (class III chitinase)	Sugarbeet (*Beta vulgaris*)	*C. nicotianae*	+
		Cucumber	*R. solani*	+
		Tobacco	*R. solani*	+
	PR5	Tobacco	*P. parasitica var nicotianae*	–
	SAR 8.2(d)	Tobacco	*Phytophthora parasitica*	+
	SAR 8.2d	Tobacco	*Pythium torulosum*	+
Tobacco (*N. sylvestris*)	PR3 (class I) chitinase	Tobacco	*Cercospora nicotianae*	–
Tomato	PR2 (class I glucanase)	Tobacco	*Fusarium oxysporum* f.sp. *lycopersici*	+
(*Lycopersicon esculentum*)	PR3 (class I chitinase)	Tobacco	*Fusarium oxysporum* f.sp. *lycopersici*	–
Wheat	PR3 (class II chitinase)	Tomato	*Verticillium dahliae*	+
(*Triticum aestivum*)	PR3 (class II chitinase)	Barley	*Erysiphe graminis*	+

Table 18.6: Two genes used in combination for making fungus-resistant transgenic plants.

Plant species	Gene 1	Donor	Gene 2	Donor	Fungus tested	Resistance
Carrot	PR3 (class I chitinase)	Tobacco	PR2 (class I glucanase)	Tobacco	*Alternaria dauci*	+
					Alternaria radicina	+
					Cercospora carotae	+
					Erisyphe heraclei	+
Tobacco	PR3 (class I chitinase)	Rice	PR2 (class II glucanase)	Alfalfa	*C. nicotianae*	+
	PR3 (class II chitinase)	Barley	PR2 (class II glucanase)	Barley	*R. solani*	+
					Alternaria alternata	+
					B. cinerea	+
	PR3 (class II chitinase)	Barley	Type I RIP	Barley	*R.solani*	+
					A. alternata	+
					B. cinerea	+
Tomato	PR3 (class I chitinase)	Tobacco	PR2 (class I glucanase)	Tobacco	*Fusarium oxysporum* f. sp. *lycopersici*	+

R genes or by manipulating genes of the SAR pathway. Diseases caused by bacterial pathogens are also covered wherever appropriate as there is considerable commonality in modes of pathogenesis and plant responses in fungal and bacterial diseases.

18.6 Transgenics with antifungal molecules

Antifungal compounds include antifungal proteins from plants and lower organisms and metabolites like phytoalexins.

18.6.1 Antifungal proteins

Till date, genes encoding many antifungal proteins which can inhibit fungal growth *in vitro* have been exploited to make fungus resistant transgenic plants although, it is not known whether they are also involved in defense responses against fungi *in vivo*. Some of these proteins are: Pathogenesis related proteins, Ribosome-inactivating proteins, Small cystein-rich proteins, Lipid transfer proteins, Storage albumins, Polygalacturonase inhibitor proteins (PGIPS), Antiviral proteins and Non-plant antifungal proteins.

18.6.2 Phytoalexins

Phytoalexins are antimicrobial and often antioxidative substances synthesised *de novo* by plants that accumulate rapidly at areas of pathogen infection. They are broad spectrum inhibitors and are chemically diverse with different types characteristic of particular plant species. Phytoalexins tend to fall into several classes including terpenoids, glycosteroids and alkaloids; however, researchers often find it convenient to extend the definition to include all phytochemicals that are part of the plants defensive arsenal.

Function of phytoalexins

Phytoalexins produced in plants act as toxins to the attacking organism. They may puncture the cell wall, delay maturation, disrupt metabolism or prevent reproduction of the pathogen in question. Their importance in plant defense is indicated by an increase in susceptibility of plant tissue to infection when phytoalexin biosynthesis is inhibited. Mutants incapable of phytoalexin production exhibit more extensive pathogen colonisation as compared to wild type. As such, host-specific pathogens capable of degrading phytoalexins are more virulent than those unable to do so.

When a plant cell recognises particles from damaged cells or particles from the pathogen, the plant launches a two-pronged resistance: a general short-term response and a delayed long-term specific response. As part of the induced resistance, the short-term response, the plant deploys reactive oxygen species such as superoxide and hydrogen peroxide to kill invading cells. In pathogen interactions, the common short-term response is the hypersensitive response, in which cells surrounding the site of infection are signalled to undergo apoptosis, or programmed cell death, in order to prevent the spread of the pathogen to the rest of the plant.

18.7 Transgenics engineered for protein

In the above section, we have discussed the development of transgenics using genes which encode for antifungal compounds like PR proteins, phytoalexins, toxins, etc. However, it appears that genes encoding these antifungal proteins provide resistance to only a limited level and to only a limited number of fungi. For example, over-expressing the chitinase gene did not provide resistance against fungi lacking chitin. Moreover, a fungus can modify its cell wall by biosynthesis of more chitosan or glucan in place of chitin and therefore, may become pathogenic again or it can evolve mechanisms to detoxify certain phytoalexins. Sexually reproducing fungi may develop resistance much faster. Furthermore, since plants are attacked by different micro-organisms during their life cycle, absence of one kind of pathogen (e.g., chitinase sensitive) will benefit other pathogens. Currently strategies that will lead to more durable and broad spectrum resistance in transgenic plants are being investigated. These strategies depend upon pathogen-induced cell death and general defense responses occurring in plants during incompatible plant – pathogen interactions.

18.7.1 Resistance genes from plants

All plants have passive defense lines such as cell walls, wax layers and chemical barriers against pathogens. If the pathogen overcomes this first line of defense, there is a second line of defense, which is mounted by proteins encoded by specific resistance (*R*) genes. This line of defense is best described genetically by the gene for gene model. It requires a pathogen protein encoded by an avirulence (*Avr*) gene to be recognised by a plant protein encoded by a resistance (*R*) gene. This activates an array of defense mechanisms, including the hypersensitive response. The gene-for-gene model although first proposed in flax-rust system, explained the genetics of resistance in other pathogens as well whether obligate or facultative.

Resistance gene-avirulence gene two-component system: De Wit proposed a model of expressing both the resistance gene (*R*) and avirulence gene (*Avr*) in the plant. When this *R–Avr* gene cassette is put under strict pathogen-inducible promoter, resistance reactions like HR will be activated upon pathogen infection.

Barnase–barstar two component system: Barnase, a cytotoxic protein with RNAse activity and barstar, its inactivator, are two proteins present in *Bacillus amyloliquefaciens*. Stritmatter and others placed the *barnase* gene under the control of pathogen-inducible potato *prp-1-1* promoter so that *barnase* activity kills the cells at the site of infection.

18.7.2   Broad spectrum disease resistance using SAR

One of the effective strategies for broad spectrum plant disease resistance has been to exploit SAR pathway. Several plant mutants have been obtained that constitutively induce SAR. Such lesion-mimic mutations have been effective in designing resistance to powdery mildew in barley. However the cell death lesions were not tightly regulated and plants were dwarfed. A major challenge is to develop transgenics that can express SAR pathway without such deleterious side effects.

Resistance responses in such transgenic plants are not constitutively activated when plants are grown under non-inducing conditions. However, upon infection by pathogen like *P. syringae* and *Peronospora parasitica* the responses are induced at higher levels. Negative regulation or mutation in genes like *MAP4* kinase of *Arabidopsis* has been shown to induce constitutive SAR response but without lesions. This opens up yet another avenue for induction of broad spectrum resistance in plants.

However finer understanding of regulation of genes involved in SAR will help us to develop resistant transgenic plants without undesired side effects like dwarfism and sterility. A significant challenge is to understand the means by which plants sense pathogens in the absence of the *R* genes. Various defense pathways can be activated by virulent pathogens which are not recognised by *R* genes, suggesting that other pathogen surveillance mechanisms exist which attenuate the severity of disease. Understanding these mechanisms will provide us with more options for developing fungus resistance in crop plants.

18.7.3   Other approaches to induce cell death

One of the earliest events in incompatible plant pathogen interaction is oxidative burst during which active oxygen species such as H_2O_2 are produced. H_2O_2 triggers production of phytoalexins, PR proteins and other HRrelated processes.

H_2O_2 also has a direct inhibitory effect on microbial growth. Glucose oxidase (GO), an enzyme occurring in some bacteria and fungi, brings about the oxidation of β-D-glucose, yielding gluconic acid and H_2O_2. GO has not been found in animals and plants. Expressing a GO gene from a fungus *Aspergillus niger* in potato showed increased level of H_2O_2. Such transgenics had reduced susceptibility to *E. carotovora* subspecies *carotovora*, *P. infestans* and *Verticillim dahliae*. Thus, molecular events occurring during plant–pathogen interactions has expanded significantly in the last ten years. Based on this, several strategies have emerged for developing crop varieties resistant to pathogens.

Strategies include the manipulation of resistance by expression of PR proteins, antifungal peptides and manipulation of biosynthesis of phytoalexins. However, in these cases the observed resistance was not absolute and was restricted to a limited number of fungi. For the antifungal compounds strategy to be successful in the long term, level of resistance in transgenic plants should be increased and its range should be broadened by isolating new genes and by testing new combinations of genes.

Genetic manipulation of the regulatory mechanisms and signalling processes controlling the coordinate activation of multiple defense responses like SAR might be the ultimate approach to modify plant resistance. However, this requires precise knowledge of both the signalling pathways involved and subsequent metabolic pathways that get triggered. While exploiting the genes in signalling pathway for making fungus resistant transgenic plants one needs to be cautious about the role of the signalling gene in various other pathways which would lead to undesirable side effects in transgenic plants. The earlier the gene function in the pathway, the greater the intricacies of regulation that will have to be addressed. Correct temporal and spatial expression of the transgene will be of critical importance and will require the availability of well-defined, pathogen-inducible promoters with the desired properties.

18.8   Virus resistant plants

Plant viruses are viruses that affect plants. Like all other viruses, plant viruses are obligate intracellular parasites that do not have the molecular machinery to replicate without a host. Plant viruses can cause damage to stems, leaves and fruits and can have a major impact on the economy because of food supply disruptions. Viruses also cause many important plant diseases and are responsible for huge losses in crop production and quality in all parts of the world. Some such diseases with their yield losses, are listed in Table 18.7.

Table 18.7: Important viral diseases of crops.

Crop	Disease	Yield loss (%)	Virus	Virus group
Cassava	Mosaic	18–25	Indian cassava mosaic virus	Begomovirus
Cotton	Leaf curl	68–71	Cotton leaf curl virus	Begomovirus
Groundnut Mungbean Blackgram	Bud necrosis	> 80	Groundnut bud necrosis virus	Tospovirus
Soyabean	Yellow mosaic	21–70	Mungbean yellow mosaic virus	Begomovirus

(Cont'd...)

Crop	Disease	Yield loss (%)	Virus	Virus group
Pigeonpea	Sterility mosaic	> 80	Pigeonpea sterility mosaic virus	Tenuivirus
Potato	Mosaic	85	Potato virus Y	Potyvirus
Rice	Rice tungro	10	Rice tungro badna and rice tungro spherical viruses	Badnavirus and waika virus
Sunflower	Necrosis	12–17	Sunflower necrosis virus	Ilarvirus
Tomato	Leaf curl	40–100	Tomato leaf curl virus	Begomovirus

Strategies for the management of viral diseases normally include control of vector population using insecticides, use of virus-free propagating material, appropriate cultural practices and use of resistant cultivars. However, each of the above methods has its own drawback. Rapid advances in the techniques of molecular biology have resulted in the cloning and sequence analysis of the genomic components of a number of plant viruses. A majority of plant viruses have a single-stranded positivesense RNA as the genome. However, some of the most important viruses in tropical countries like India have single-stranded and double-stranded DNA genomes and RNA genomes of ambisence polarity, i.e., genes oriented in both directions.

Concomitantly, tremendous advances have taken place of plant virus interaction in the process of pathogenesis and resistance. This, along with associated advances in the genetic transformation of a number of crop plants, have opened up the possibility of an entirely new approach of genetic engineering towards controlling plant virus diseases.

There are mainly two approaches for developing genetically engineered resistance depending on the source of the genes used. The genes can be either from the pathogenic virus itself or from any other source. The former approach is based on the concept of pathogen-derived resistance (PDR). For PDR, a part, or a complete viral gene is introduced into the plant, which, subsequently, interferes with one or more essential steps in the life cycle of the virus. This was first illustrated in tobacco by the group of Roger Beachy, who introduced the coat protein (CP) of tobacco mosaic virus (TMV) into tobacco and observed TMV resistance in the transgenic plants. The concept of PDR has generated lot of interest and today there are several host–virus systems in which it has been fully established. Non-pathogen-derived resistance, on the other hand, is based on utilising host resistance genes and other genes responsible for adaptive host processes, elicited in response to pathogen attack, to obtain transgenics resistant to the virus.

18.9 Transgenics with pathogen-derived resistance

In a number of crops, transgenics resistant to an infective virus have been developed by introducing a sequence of the viral genome in the target crop by genetic transformation. Virus-resistant transgenics have been developed in many crops by introducing either viral CP or replicase gene encoding sequences.

Resistance obtained by using CP is conventionally called CPMR. Replicase-mediated resistance has been pursued in a number of laboratories and in most of these cases, resistance has been shown to be due to an inherent plant response, known as post-transcriptional gene silencing (PTGS). Because of the essential nature of the viral movement protein for intercellular movement of plant viruses, movement problem sequence has also been used for achieving viral resistance.

18.9.1 Coat protein

Phage major coat protein is an alpha-helical protein that forms a viral envelope of filamentous bacteriophages. These bacteriophages are flexible rods with a helical shell of protein subunits surrounding a DNA core. The approximately 50-residue subunit of the major coat protein is largely alpha-helix and the axis of the alpha-helix makes a small angle with the axis of the virion.

The use of viral CP as a transgene for producing virus resistant plants is one of the most spectacular successes achieved in plant biotechnology. Numerous crops have been transformed to express viral CP and have been reported to show high levels of resistance in comparison to untransformed plants (Tables 18.8 and 18.9). Powell-Abel and others first reported resistance against TMV in transgenic tobacco expressing the TMV CP gene. The resistance was manifested as delayed appearance of symptoms as well as a reduced titre of virus in the infected transgenic plants, as compared to the controls.

Table 18.8: Coat protein-mediated transgenic resistance to viruses in crops.

Crops	*Viruses**	*Field tested***
Cereals		
Maize	MDMV, MCMV	n.r.
Rice	RSV, RTSV	n.r.
Wheat	WSMV	n.r.
Fruits		
Apricot	PPV	n.r.
Cantaloupe	ZYMV, WMV2, CMV	Yes
Citrus	CTV	n.r.
Grape	GCMV, GFLV, ToRSV	n.r.
Muskmelon	ZYMV	Yes

(Cont'd...)

Crops	Viruses*	Field tested**
Papaya	PRV	Yes
Plum	PPV	n.r.
Squash	ZYMV,WMV2	Yes
Vegetables		
Pepper	TSWV	n.r.
Tomato	ToMV, YMV, CMV, TYLCV	Yes
Potato	PVX, PVY, PLRV	Yes
Lettuce	LMV, TSWV	n.r.
Pea	PEMV	n.r.
Cucumber	CMV	Yes
Sugarbeet	BNYVV	n.r.
Legumes		
Peanut	TSWV	n.r.
Soyabean	BPMV	n.r.
Bean	BPMV	n.r.

*MCMV, Maize chlorotic mottle virus; MDMV, Maize dwarf mosaic virus; RSV, Rice stripe virus; RTSV, Rice tungro spherical virus; WSMV, Wheat streak mosaic virus; CTV, Citrus tristeza virus; GCMV, Grapevine chrome mosaic virus; GFLV, Grapevine fanleaf virus; ToRSV, Tomato ringspot virus; YMV, Yellow mosaic virus; LMV, Lettuce mosaic virus; PEMV, Pea enation mosaic virus; BNYVV, Bean necrotic yellow vein virus; BPMV, Bean pod mottle virus **n.r. indicates not reported.

Table 18.9: Comparative performance of transgenic virus resistant plants.

Host	Transgene	Yield increase (%)
Tomato	TMV CP	40
Tomato	CMV satellite	14
Potato	PVX + PVY CP	38
Squash	CMV + ZYMV + WMV2 CP	97
Squash	ZYMV + WMV2 CP	90
Squash	ZYMV CP	77
Papaya	PRSV CP	90

The resistance against TMV using TMV CP in tobacco was also reported to be effective against other tobamoviruses whose CP was closely related to that of TMV but not effective against viruses which were distantly related to TMV. Transgenic potato, expressing the CP of potato virus X (PVX) also showed resistance against PVX.

However, in marked contrast to TMV, this resistance was not broken down when PVX RNA was used as the inoculum, thus indicating several possible mechanisms of CPMR.

Replicase (Rep)

Replicase (Rep) protein-mediated resistance against a virus in transgenic plants was first shown in tobacco against TMV in plants containing the 54 kDa putative *Rep* gene. Similar resistances have been developed for several other viruses namely pea early browning virus, PVY and CMV.

Gene constructs of *Rep* genes that have been used for resistance include full-length, truncated or mutated genes. Many of the above resistance responses have now been shown not to require protein synthesis and to be mediated at the RNA level, which is described in more detail later under 'post-transcriptional gene silencing'.

This type of resistance remains confined only to a narrow spectrum of viruses, the spectrum being narrower than that of CPMR. To make the resistance broad-based, it may be necessary to pyramid such genes from several dissimilar virus sources into the test plant genome. However, the resistance generated by the use of Rep sequences is very tight; a high dosage of input virus can be resisted easily by the transgenic plant.

Movement protein

Successful infection of a plant by a plant virus depends on its ability to move from the cell initially infected to neighbouring cells in order to spread infection. Unlike animal cells, plant cells have robust cell walls, which viruses cannot easily penetrate. A movement protein is a non-structural protein which is encoded by some plant viruses to enable their movement from one infected cell to neighbouring cells.

Many, if not all, plant viruses encode a movement protein and some express more than one. The movement protein of tobacco mosaic virus (TMV) has been most extensively studied. Plant viruses can also be transported over longer distances through the host plant in the vascular system, via., the phloem.

Satellite RNA

Plant viruses often contain parasites of their own, referred to as satellites. Satellite RNAs are dependent on their associated virus for both replication and encapsidation. Satellite RNAs vary from 194 to approximately 1500 nucleotides (nt). The larger satellites (900 to 1500 nt) contain open reading frames and express proteins *in vitro* and *in vivo*, whereas the smaller satellites (194 to 700 nt) do not appear to produce functional proteins. The smaller satellites contain a high degree of secondary structure involving 49 to 73% of their sequences, with the circular satellites containing more base pairing than the linear satellites. Many of the smaller satellites produce multimeric forms during replication.

Some of these smaller satellites encode ribozymes and are able to undergo autocatalytic cleavage. The enzymology of satellite replication is poorly understood, as is the replication of their helper viruses. In many cases the coreplication of satellites suppresses the replication of the helper virus genome. This is usually paralleled by a reduction in the disease induced by the helper virus; however, there are notable exceptions in which the satellite exacerbates the pathogenicity of the helper virus, albeit on only a limited number of hosts. The ameliorative satellites are being assessed as biocontrol agents of virus-induced disease. In greenhouse studies, satellites have been known to 'spontaneously' appear in virus cultures.

Defective-interfering viral nucleic acids

In several viruses, truncated genomic components are often detectable in infected tissues, which interfere with the replication of the genomic components. These species of DNA are also called defective interfering (DI) DNA and expression of delayed disease symptoms and recovery, coupled with increased resistance upon repeated inoculation have been observed in plants engineered with DI DNA. For example, incorporation of subgenomic DNA B that interferes with the replication of full length genomic DNA A and B confers resistance to ACMV in *N. benthamiana*.

18.9.2 Transgenics with non-pathogen derived resistance

The following section describes the non-pathogen-derived strategies, i.e., those utilising genes derived from either the host plant or any other non-pathogenic source. A new phenomenon called post-transcriptional gene silencing (PTGS) has recently been shown to be responsible for the inherent ability of many plants to specifically degrade nucleic acids in a sequence-specific manner, including those of viruses. Thus, this strategy can be very effective in engineering virus resistance. The other nonpathogen derived strategies are the utilisation of plant disease resistance genes, the ribosome-inactivating proteins, plant proteinase inhibitors, human interferon-like systems, antiviral antibodies expressed in plants, systemic acquired resistance and secondary metabolite engineering.

Post-transcriptional gene silencing

Post-transcriptional gene silencing (PTGS) is a specific RNA degradation mechanism of any organism that takes care of aberrant, unwanted excess or foreign RNA intracellularly in a homology-dependent manner. It is prevalent in various forms of life, namely plant, fungus and invertebrate animals. This activity could be present constitutively to help normal development or induced in response to cellular defense against pathogens. In this mechanism, the elicitor double-stranded RNA (ds RNA), commonly produced during viral infection, is degraded to 21–25 nucleotides, termed as small interfering RNA (siRNA), with the help of a variety of factors that have already been or are being identified.

Plant disease resistance genes

A number of disease resistance genes (R) have been reported against viruses of crop plants. They encode products which respond to viral signals (avirulence (*avr*) gene products) culminating in a number of resistance responses in the plant. As shown in Table 18.10, many of the corresponding viral *avr* genes have also been identified. Some of the R genes have been shown to complement the disease susceptibility phenotype in the corresponding cultivars when used as transgenes, furnishing a direct proof of their action. The following section describes the current knowledge about R genes against viruses and their mechanisms of action.

Table 18.10: R genes against viruses and corresponding *avr* gene products.

Resistance gene	Source plant	Avr product of the virus	Pathogen
HRT	*Arabidopsis thaliana* ecotype Dijon	Coat protein	TCV
I	*Phaseolus vulgaris*	n.d.	BCMV
L2	*Capsicum* sp.,	Coat protein	PMMV
L3	*Capsicum* sp.,	Coat protein	PMMV
N	*N.tabacum* cultivar *Samsun*	Replicase	TMV
RRT	*Arabidopsis thaliana* ecotype Dijon	Coat protein	TCV
RTM	*Arabidopsis thaliana* ecotype Columbia-O	n.d.	TCV
Rx, Nx, Nb	*Solanum tuberosum* cultivar *Cara*	Coat protein	PVX
Ry	*Solanum stoloniferum*	NIa protease	PVY
Tm1	*Lycopersicon esculentum*	Replicase	TMV
Tm2	*L. esculentum*	Movement protein	TMV
Tm2(2)	*L. esculentum*	Movement protein	TMV
TuRB01	*Brassica napus*	Cylindrical Inclusion protein	TuMV
Va	*Nicotiana tabacum* cultivar *Burley*	Covalently-linked viral genomic protein	TVMV

n.d. not determined; PMMV, Pepper mild mosaic virus; BCMV, Bean common mosaic virus; TVMV, Tobacco vein mottling virus.

The other anti-viral R genes which have been identified are *Sw-5* and *Tsw* against TSWV from tomato and pepper respectively, *Ry* against PVY, from *Solanum stoloniferum*, *Va* against tobacco vein mottling virus (TVMV) from

N. tabacum cultivar Burley, *TuRB01* from *Brassica napus* against TuMV, *I* against bean common mosaic virus (BCMV) from *Phaseolus vulgaris*, *L2* and *L3* against pepper mild mottle virus (PMMV) from *Capsicum* sp., *Nx* and *Nb* against PVX from *Solanum tuberosum* and *Tm1*, *Tm2* and *Tm2(2)* against TMV from *Lycopersicon esculentum*.

The most exciting approach towards engineering improved resistance to multiple diseases may be the development of new R genes having multiple specificities. The *Fen* (resistance to the insecticide Fenthion) and *Pto* genes are located in the same R gene cluster in the tomato genome and they are 86% identical in nucleotide sequence. A functional gene was made by domain swapping of the two genes, thus raising the possibility of creating a hybrid gene containing multiple specificities. Another novel strategy, termed two-component approach, has been developed lately and holds lot of promise for introducing broad-spectrum resistance.

Ribosomal inactivating proteins

Several plants have been found to contain antiviral proteins, commonly termed as ribosome-inactivating proteins (RIPs). RIPs inhibit the translocation step of translation by catalytically removing a specific adenine base from 28S ribosomal RNA. They are synthesised either as pre-or pre-pro-proteins and targeted to vacuoles. Because of their specific intracellular localisation, RIPs do not affect the endogenous 28S RNA. It is supposed that RIPs enter cells together with the viruses and exert the damage to the host ribosome or possibly viral RNA. The antiviral activity of several types of RIPs has been well-documented. When purified RIPs are mixed with viruses and applied on plants, virus multiplication and symptom development are dramatically suppressed. A broad range of viruses can be suppressed in this manner.

Protease inhibitors from plants

Many viruses, namely poty-, tymo-, nepo-, como- and closteroviruses need cysteine protease activity to process their own polyproteins for their replication and propagation. Hence plants expressing cysteine protease inhibitors might resist the growth of viruses as mentioned above. This idea was tested by using cysteine protease inhibitors (oryzacystatin) of rice to successfully engineer resistance against potyviruses in transgenic tobacco plants.

Tobacco lines expressing the rice cysteine proteaseinhibitor gene were examined for resistance against tobacco etch virus (TEV) and PVY infection.

Interferon-like systems

Higher vertebrates resist virus infections in part by catalysis of RNA decay using the interferon regulated 2–5A system. The 2–5A system consists of two enzymes, namely a 2–5A synthetase that makes 5′ phosphorylated, 2′-5′- linked oligoadenylates (2–5A) in response to doublestranded DNA and the 2–5A dependent RNAse L. In plants, homologues of this system are not yet known but the inducers, i.e., interferon-like molecules have been reported.

Anti-viral plantibodies

Another approach to control plant viruses is to express specific anti-viral antibodies in plants, commonly known as plantibodies. The efficacy of this approach has been demonstrated against Artichoke mottled crinkle virus in transgenic *N. benthamiana*. A panel of monoclonal antibodies was raised against AMCV and the gene for the most reactive of the above panel was cloned and expressed in *N. benthamiana*. The above transgenic plants and their progeny showed lower virus accumulation, reduced incidence of infection and delayed symptom appearance, compared to non-transgenic plants.

Systemic acquired resistance

Following viral infections, plants develop an active resistance which is at first localised only at the site of infection, but spreads systemically in due course. This resistance, called systemic acquired resistance (SAR), is characterised by the coordinate activation of several genes in uninfected, distal parts of the inoculated plants. SAR is characteristically associated with accumulation of salicylic acid (SA), enhanced expression of pathogenesisrelated (PR) proteins activation of phenylpropanoid pathway, leading to the synthesis of higher phenolic compounds, increase of active oxygen species and reinforcement of cell wall by the deposition of lignin and suberin. Involvement of SA in TMV resistance has been shown by expressing the bacterial salicylate hydroxylase (*NahG*) gene in tobacco plant, thus decreasing its endogenous salicylic acid and causing susceptibility to TMV infection.

The discovery that SA-binding protein is a catalase, whose activity is blocked by SA led to the proposal that the mode of action of SA is to inhibit the hydrogen peroxide degrading enzyme catalase, resulting in elevation of hydrogen peroxide levels. Transgenic tobacco plants were developed that expressed catalase 1 (*Cat1*) or catalase 2 (*Cat2*) gene in an antisense orientation.

Antisense catalase transgenic plants exhibiting severe reduction in catalase activity (approximately 90% or more), developed chlorosis or necrosis on lower leaves. These plants also showed high level of SA and PR accumulation as well as enhanced resistance to TMV.

Secondary metabolite pathways

Metabolic pathways which are important in viral pathogenesis are key targets for intervention against viral infection. One such step is mediated by *S*-adenosyl homocystein hydrolase (SAHH), which is a key enzyme in trans-methylation reactions that take place, using Sadenosylmethionine as the methyl donor. It is suggested to play a role in 5' capping of mRNA during replication. The antisense RNA for tobacco SAHH was expressed in transgenic tobacco plants. Though 50% of the plants showed stunting, they were resistant to infection by various plant viruses.

18.9.3   Essential considerations for developing virus-resistant transgenics

Variability

Viral genes show high levels of variability. This may be due to lack of proof reading function of viral replicases and the high recombination rates of viral genomes during the progress of infection. Symptomatic variants or strains of viruses, as well as geographically distinct isolates, not showing such variations in symptoms, have been nevertheless, documented to contain significant variability in their genes. Under field conditions, most of the viruses are believed to exist as collection of variants, or 'quasispecies', as documented in cassava-infecting geminiviruses in Uganda and rice tungro bacilliform virus, a double-stranded DNA virus in southeast Asia.

As with naturally occurring virus resistance genes, when considering virus resistance under field conditions, strain specificity and breadth of protection are important questions. There is often a general correlation between the extent of protection and the relatedness between the challenge virus and virus from which the transgene was derived. It is clear from the case of transgenic papaya that the level of resistance is dependent upon the homology between the prevalent viral isolate and the transgene. It is imperative that in any viral transgene strategy, sequence of the aggressive prevalent strain of the virus in that region is used.

Sufficient information on the degree of diversity amongst the biologically indistinguishable viral strains needs to be collected before designing the transgene. It is especially true of whitefly transmitted geminiviruses, where the evolution of the virus is rapid. A wide variety of virus genotypes may be present, either maintained in different cultivated hosts or on endogenous weed species. Depending upon change in the vector behaviour, e.g., feeding on to a new host more frequently than it was doing earlier and vector population build-up, viruses of different populations may start infecting new hosts leading to further changes in their genotype.

The success of any transgenic strategy is dependent upon the level of resistance to multiple inoculation of the same or related strains, by vector transmission. In recent years, efforts have been made to identify the variants and to assess the genetic relatedness between them. However, frequency distribution of these variants in a given virus population needs to be assessed to develop a transgenic strategy targeting any virus causing an economically important disease. The population structure of the virus is determined by evolutionary factors affecting its life cycle, the major factor being selection pressure on the gene products that interact with host and the vector. Variability may result due to host component as new host genotypes are introduced, or by vector component as they adapt to new host system or by the virus itself by mutation, complementation or recombination. A periodical assessment of population structure is mandatory if virus-derived transgenic resistance strategy is adopted for the control of the disease.

Biological risks

The concept of using pathogen-derived genes to induce transgenic resistance has no doubt raised a number of ecological concerns. Risk perceptions boil down to two major items: (i) recombination between viral-derived transgene and non target virus, (ii) transmission/vector host range changes brought about by heteroencapsidation, i.e., encapsidation of the genome of non-target virus with the transgenically expressed CP.

Field trials conducted so far with transgenics have not indicated that expression of viral transgenes leads to the emergence of new super strain or change in transmission behaviour of common viral pathogens. However, sufficient care should be taken to avoid any risks due to heteroencapsidation while designing the constructs.

The strong linkages shown by CP with insect transmission of viruses, have made possible heteroencapsidation, an important factor to be considered while designing CPbased transgenes. Coat protein genes have been designed from PPV, such that a 'DAG' motif in the CP, believed to play an important role in vector transmission, was deleted to prevent any further insect transmission of heteroencapsidated virions.

The use of these constructs in producing transgenic plants has shown that heteroencapsidation of ZYMV was significantly reduced without compromising virus resistance of the plants. Similar results have also been reported recently in transgenic *N. benthamiana* expressing mutated PPV CP, which were not only resistant to PPV, but were also suppressed in heteroencapsidation, when infected with chilli vein mottle virus and PVY.

Comparison of anti-viral strategies

The success of transgenic approach varies for any specific host/virus combination. A range of phenotypes is observed amongst the virus-resistant transgenic plants. While CPMR confers broad-spectrum, less complete resistance, Rep-mediated resistance produces immunity against the virus, but to a limited spectrum of strains. Similarly, in RNA-mediated resistance, antisense RNA targeting mRNA of DNA viruses has more potential than against positive stranded RNA virus. Any antisense RNA/ribozyme strategy should bear in mind the association/ dissociation parameters of the molecules. Pyramiding of different transgenes or combination of transgenes with natural resistance targeting different events in viral life cycle will increase the confidence level in the management of viral diseases and will ensure stability of resistance at the field level. Durability, broad-spectrum character of the transgene-derived resistance coupled with enhanced crop yield of the transgenics *viv-à-vis* healthy, untransfomed plants, etc., are some of the essential parameters, which any important strategy must incorporate.

Engineering plant quality and proteins

19.1   Introduction

Plants are the main source of micro- and macro-nutrients for most living organisms. Their capacity to photosynthesise allows them to transform solar energy into chemical energy stored in organic and inorganic molecules and use it for their own necessities. Animals, including humans, need plant nutrients, since many nutrients that are essential for the correct performance of vital animal functions are only synthesised by plants. For millions of years, humans have selected outstanding characteristics of crop plants like vigour, yield, flavour, resistance to pests and higher nutritional value and now-a-days the major cereals (rice, wheat, maize, barley, millet and sorghum) domesticated from wild relatives are the main source for human and animal nutrition. This selection has been made, based mainly on phenotypic characteristics and successful progress has been some what slow. However, the human population has increased rapidly within the last hundred years and the need for appropriate foods, especially in developing countries is a priority for producers. The development of molecular biology offers the possibility for improving the desired characteristics of crop plants through the direct introduction of specific genes, with the advantage of being faster and more selective. Molecular biology also provides the tools for a complete molecular and physiological analysis of the transformed plants. Genetic engineering is simply an additional tool for improving crop plants in combination with traditional breeding. Both strategies have been used in recent years to increase the nutrient content of some crop plants ensuring better quality foods for certain populations. This chapter presents a general view of the biotechnological methods currently used for improving plants and how they are being exploited to increase the content of specific nutrients in specific plant species important to the human population.

19.2   Nutrition process

Nutrition is the process through which organisms acquire and process chemical substances called nutrients to develop their basic vital functions, such as growth and reproduction. These nutrients may be obtained from air, water, soil or other organisms. Plants and cyanobacteria capture the energy of sunlight and together with water and CO_2, transform it to chemical energy by means of a succession of reactions collectively named photosynthesis. The final products of photosynthesis are energy-rich sugar molecules that constitute the substrates and energy source of a number of chemical reactions that enable plants to synthesise complex molecules, from metabolites to building blocks. Along with H and C (from H_2O and CO_2), another 15 elements are believed to be essential for the growth of most cultivated plant species. An element is judged essential if in its absence the plant cannot complete its life cycle. All 17 elements are needed in different amounts. Trace elements or micro-nutrients (Mo, Ni, Cu, Zn, Mn, B, Fe, Cl) are needed in tissue concentrations equal or less than 100 mg kg^{-1} of dry matter, whereas macro-nutrients (S, P, Mg, Ca, K, N, O, C, H) are needed in concentrations of 1000 mg kg^{-1} of dry matter.

Since animals are unable to photosynthesise, they consume and process energy rich molecules from other organisms to obtain their nutrients. In order to maintain optimal health, animals including humans require the consumption of a set of macro and micro-nutrients. Animals therefore, depend on photosynthetic organisms either directly or indirectly to support their nutrition. Within this panorama, photosynthetic organisms are considered the chemical source of energy for life on earth and plants, as photosynthetic organisms are essential for the survival of other non-photosynthetic species.

Animal macro-nutrients (carbohydrates, lipids and proteins) make up most of the bulk of foodstuff and are used primarily as an energy supply, whereas micro-nutrients (organic and inorganic compounds) are needed in small amounts and contrary to macro-nutrients, are not used as energy sources. Nonetheless, micro-nutrients are essential components of cells and tissues and are also required to accomplish physiological functions, such as muscle contraction and the conduction of nerve impulses. Essential micro-nutrients in the human diet include 17 minerals and 13 vitamins. Non-essential micro-nutrients encompass a vast group of unique organic phytochemicals that are not strictly required in the diet, but are linked to the promotion of good health. To prevent nutrient deficiencies in humans, each group of nutrients has to be consumed at a daily intake level above a defined minimum value. The level of requirement of each nutrient varies with age, sex and physiological status. Additionally, the intake of some nutrients at higher, therapeutic levels has been associated with a reduction in risk for chronic conditions.

Plant components with nutritional significance or benefit for human health have been named phytonutrients. These plant-derived nutrients are not ready packaged in specific plants to satisfy human needs, they are distributed throughout a variety of plant species to satisfy the metabolic requirements of each species, furthermore each plant species is found naturally in a specific environment and not distributed worldwide. The perfect plant, to satisfy human needs in nutritional terms, does not exist. Moreover, modern agriculture has become highly specialised concentrating on a handful of staple crops namely:

rice, maize, wheat, potatoes, sweet potatoes and grain legumes. Cereal crops are rich in carbohydrates, providing more than 80% of calorie intake and grain legumes rich in protein often replace meat in the diets of populations in developing countries. However, these crops are poor sources of some macro-nutrients and many micro-nutrients, which can lead to serious health problems in the human population. For this reason, traditional breeding efforts and more recently biotechnological strategies have focused on increasing the content of such nutrients in staple crops.

19.3   Crop improvement

Plant breeding has been the main strategy to improve the nutritional value of plants. Traditional plant breeding strategies are based on the knowledge obtained through the investigations of the inheritance of any character, the mode of inheritance and existing genetic variability. Appropriate breeding strategies make use of phenotypic selection of individuals or families. The methods for selection first originated from the crossing between individuals of the same species or closely related species. The outstanding individuals (hybrids) are selected throughout several crossing and back crossing cycles to finally obtain a generation carrying the desirable trait and which is also commercially acceptable: The new individuals are known as a new variety.

Traditional breeding has been enormously successful in producing new varieties adapted to different stresses, such as attack by pathogens and pests, or giving high yields. Unfortunately, the majority of breeding efforts are aimed at the needs of consumers in developed countries and not at improving crops for use in developing countries where environmental conditions, diseases and pests can often have devastating effects.

Despite their success, traditional breeding methods do have certain drawbacks. Often desirable traits are found in land races or other closely related species, therefore to introduce the character of interest, many rounds of subsequent selection are necessary to reduce the linkage drag or remove all the undesirable traits present in the non-commercial parent. Another drawback is the effect of the environment on the trait of interest. Therefore, breeders must distinguish between the real genetic potential of the plant and the effect the environment may have. Finally, simply producing inter-species crosses can be difficult, since they produce few viable seeds from which to select.

In part, these drawbacks are being alleviated by the increasing use of molecular markers associated with traits of interest. Selection can be made on the basis of molecular marker genotype eliminating the problem of environmental effects and allowing selection on a genome-wide basis. Therefore, selection for the elimination of undesirable traits is more effective. Molecular markers are also important for the selection of complex traits where multiple genes are involved.

As will be discussed below, marker assisted selection for complex traits such as yield, drought tolerance, etc., will be an essential tool in producing new varieties at least until the underlying genetic mechanisms governing these traits have been revealed.

19.3.1 Genetic engineering

Genetic engineering refers to sophisticated, artificial techniques capable of transferring genes from unrelated organisms directly to recipient organisms. Genes are the units of genetic information found in all organisms and they contain regulatory elements that stimulate or silence their expression in a tissue and temporal specific manner. Genes determine specific traits, like colour, height, or tolerance to frost. Adding novel genes to crops means adding new traits and abilities. Genetic engineering allows the incorporation in plants of genetic information from any other organism, including bacteria, animals, insects, etc., in addition to plants.

The expression of different genes of an individual is coordinated by genetic regulatory elements (promoters, enhancers and silencers) that determine changes in growth, development and cellular differentiation. Genetic engineering leads to direct access and manipulation of the information contained in DNA. It also permits the creation of synthetic genes.

Major advances in the transformation of plant species have come from the development and improvement of transformation technologies, plant tissue culture and regeneration from transformed cells or tissues, as well as basic research in plant molecular biology and physiology. The modification of plant components using molecular genetic technology is aimed at increasing plant production, quality or quantity of products, lowering production costs and even the production of important biological molecules difficult to find or not found in plants or even in nature. Agricultural biotechnology, through genetic engineering, is making rapid progress and transgenic plants are being introduced commercially at an increasing rate. Genetic engineering reduces the time taken to obtain an improved cultivar (Fig. 19.1).

19.3.2 Identification of genes with potential to improve the nutritional quality of plants

Techniques of molecular biology are employed to study the genes of different organisms. These strategies permit the study of specific components within the genome and the interrelationship of such components as a whole. Using these means, it has been possible for scientists to identify the gene or genes whose products participate in diverse metabolic pathways. The basis for genome analysis in plants is a genomic map, which can be either a genetic map based on information from both visible and molecular markers, or a physical map, in

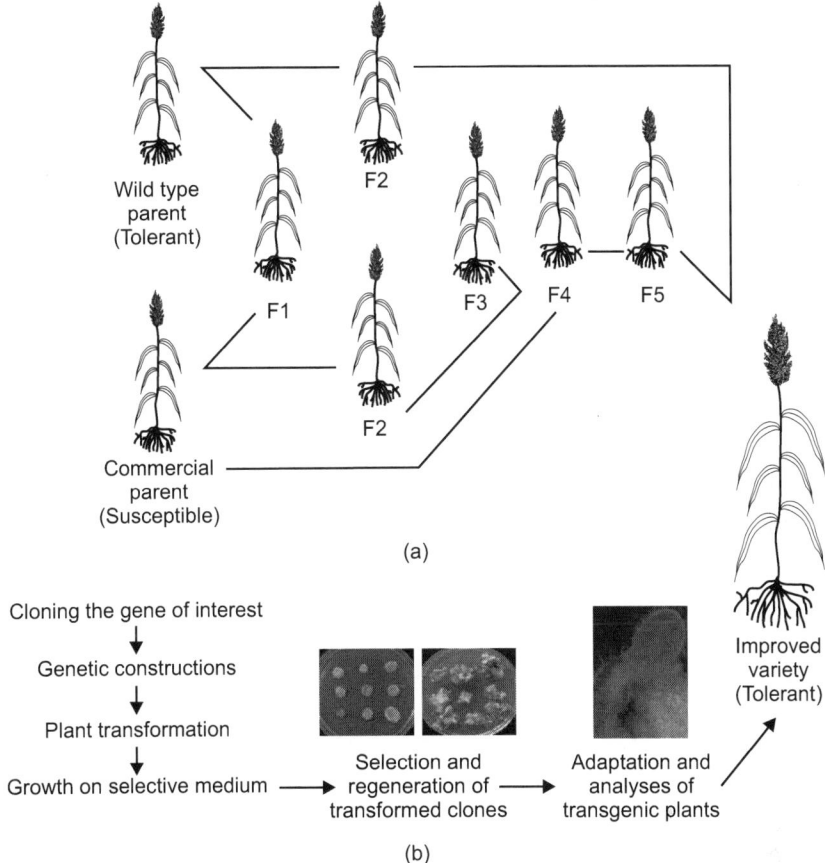

Wild type
parent
(Tolerant)

F2

F1

F3 F4 F5

F2

Commercial
parent
(Susceptible)

(a)

Improved
variety
(Tolerant)

Cloning the gene of interest
↓
Genetic constructions
↓
Plant transformation
↓
Growth on selective medium ⟶ Selection and
regeneration of ⟶
transformed clones
Adaptation and
analyses of
transgenic plants

(b)

Figure 19.1: Strategies for the improvement of crops: (a) through traditional breeding and (b) genetic engineering.

which yeast artificial chromosomes (YACs) and bacterial artificial chromosomes (BACs) are aligned with the chromosomes to give a position to genes within the genome.

A current approach to gene discovery that is most applicable to compounds of nutritional importance, synthesised or accumulated by plants and other organisms has been named nutritional genomics. This discipline takes advantage of the homologies or similarities between the metabolic routes that lead to the generation of a given product in different organisms. In this context, previously characterised genes from the metabolic pathway under research, either from animals, micro-organisms or any plant species can be found in public databases and may be used as a source of genetic information to study or modify the characteristics of the target plant. These modifications can be accomplished by the introduction of foreign genes isolated from other species or by cloning genes from the recipient plant using as a molecular tool the information from the genes already isolated from other organisms.

An example of the application of nutritional genomics is the discovery and modification of genes involved in the synthesis of vitamin E in plants. The first gene involved in this pathway in *Arabidopsis thaliana* was isolated with fungal and human orthologs as database queries. The sequence of the *Arabidopsis* gene served to identify an ortholog in a 10-gene operon in the cyanobacterium *Synechocystis* PCC6803. Gene disruption experiments showed that this operon also encodes for a γ-tocopherol methyl-transferase (γ-TMT), which is the final step in vitamin E synthesis. This γ-TMT gene of Synechocystis allowed the isolation of an ortholog from the *Arabidopsis* database, whose over-expression increased vitamin E levels nine-fold in *Arabidopsis* seed oil.

When there is a lack of information on the specific gene or genes whose products are involved in the metabolic pathway of interest, the isolation of genes becomes more complicated, but researchers have developed alternative strategies. Some of these strategies have their basis in mutagenesis, or alterations of the genome that produce changes in the phenotype due to the modification of the components of a given metabolic route.

The gene(s) modified through mutagenesis may be localised by molecular marker mapping techniques and may be further isolated and characterised. In maize, different mutants affected in seed development have been obtained and the altered genes responsible for their phenotype have been cloned using mapping techniques. Such is the case of miniature-1, whose phenotype is grain size reduction, which is the result of the lack of extracellular invertase activity. A mutant named shrunken presents starchless grains and the affected gene was identified as the endosperm sucrose synthase. Other advances have been made in the application of techniques to identify and isolate genes, including the sequencing of entire genomes of prokaryotic organisms. Recently, new approaches have been developed that allow the mutagenesis of large numbers of genes.

These procedures are of two types: first, 'random' insertional mutagenesis, in which insertion mutations are randomly generated throughout the genome, followed by the identification of the gene(s) affected by comparing the sequence adjacent to the insertions with the genome sequence, or expressed sequence tags (EST) and second, targeted mutagenesis in which specific genes are deleted or analysed. The use of transposons and retrotransposons has made it possible to carry out non-site directed mutagenesis. This approach is based on the features of natural transposable elements that are ubiquitously found in eukaryotic organisms and whose integration into a new location within the host genome can disrupt genes, effectively producing a 'tagged' mutation. A similar approach is to use *Agrobacterium tumefaciens* T-DNA insertions to produce mutations.

The methods described above generally work on a 'one gene in one experiment' basis, where the whole picture of gene function is hard to obtain. Nevertheless, other methods let scientists monitor a wider range of gene expression and therefore form a more complete picture of gene expression and the interaction of gene products. Two-dimensional electrophoresis and methods for large-scale analysis of proteome variations has evolved into 'proteomics', which can be used to map translated genes and loci controlling their expression, identifying variations of complex phenotypic traits.

Among other techniques, differential display is based on the analysis of differential mRNA populations produced in an organism as a result of its exposure to two or more different environmental conditions. In addition, large-scale DNA sequencing and microarrays are new technologies that promise to monitor the whole genome on a single chip, so that the interactions between thousands of genes can be analysed simultaneously. These advanced methods can be subjected to a combined bioinformatic based and expression-based analysis to identify a limited set of candidate genes for a specific metabolic pathway.

19.3.3   Introduction of foreign genes into plant cells

Norman Borlaug, once said 'Genetically modified organisms are the result of a natural process that was going on long before humans became involved'. The foundations for recombinant DNA technology, based on genetic manipulation, were established when enzymes which modify DNA, such as endonucleases, ligases and polymerases, were discovered and were used by scientists to isolate, characterise and modify genes and ultimately transfer them to the same organism, or a different organism, to study and exploit their expression.

The development of transgenic plants involves the following steps:

1. Identification of a gene that could impart a useful character to the target crop plant and subsequent cloning of the gene. Strategies for this have been discussed above.

2. Modification of the target gene for expression in crop plants. The gene has to be isolated and cloned into a vector, which ensures the stable integration of the foreign gene into the chromosomes of the recipient plant cells. The gene must also be placed under the control of the appropriate regulatory sequences.

3. Transfer of the modified gene into cells of the plant species.

4. Regeneration of complete transgenic plants capable of transmitting the incorporated gene to the next generation. This process involves the use of selectable markers, such as antibiotic resistance, in order to select for transformed tissue. The transformed nature of the tissue is confirmed by PCR, Southern and Western blots or expression assays.

5. Phenotypic analysis of transgenic progeny to measure expression and functionality of the transgene in the transformed plants and segregation analysis of transformed progeny.

6. Field trials of transgenic plants.

Two classes of plant transformation technologies have been developed: *In vitro* methods and 'natural' methods. Among the *in vitro* technologies are microinjection (not commonly used), direct DNA uptake into protoplasts (with or without electroporation) and microprojectile (or particle) bombardment. 'Natural' technologies include the use of viral vectors (which lead to transient unstable transformation) and *Agrobacterium tumefaciens* mediated transformation. Each of these technologies has advantages and disadvantages.

The *in vitro* techniques tend to result in transformed plants containing a high copy number of often rearranged or catenated transgenes, which can result in homology dependent suppression (transgene silencing). In addition, electroporation technology depends upon the ability to regenerate the transgenic protoplasts into whole plants, a process that is either difficult or impossible for many plant species. Particle bombardment can be performed with any tissue of most species; however, the process is relatively inefficient in that few cells are stably transformed.

Agrobacterium transformation of most dicotyledonous species is simple and efficient and methodology has been adapted to allow many monocotyledonous plants, including cereals, to be transformed. There remain, however, several important crop plants such as soyabean and common bean for which efficient transformation systems are lacking.

Transgenic animals

20.1 Introduction

Transgenic animals are animals (most commonly mice) that have had a foreign gene deliberately inserted into their genome. Such animals are most commonly created by the micro-injection of DNA into the pronuclei of a fertilised egg which is subsequently implanted into the oviduct of a pseudopregnant surrogate mother. This results in the recipient animal giving birth to genetically modified offspring. The progeny are then bred with other transgenic offspring to establish a transgenic line. Transgenic animals can also be created by inserting DNA into embryonic stem cells which are then micro-injected into an embryo which has developed for five or six days after fertilisation, or infecting an embryo with viruses that carry a DNA of interest. This final method is commonly used to manipulate a single gene, in most cases this involves removing or 'knocking out' a target gene. The end result is what is known as a 'knockout' animal.

A transgenic animal is one whose genome has been altered by the transfer of a gene or genes from another species or breed. Transgenic animals are routinely used in the laboratory as models in biomedical research. Over 95% of those used are genetically modified rodents, predominantly mice. They are important tools for researching human disease, being used to understand gene function in the context of disease susceptibility, progression and to determine responses to a therapeutic intervention. Mice have also been genetically modified to naturally produce human antibodies for use as therapeutics.

The dependence of man on animals such as cattle, sheep, poultry, pig and fish for various purposes (milk, meat, eggs, wool, etc.), is well known.

Improvement in the genetic characteristics of livestock and other domestic animals (e.g., high milk yield, weight gain, etc.), in the early days, was carried out by selective breeding methods.

This technique primarily involves a combination of mating and selection of animals with improved genetic traits. Although selective breeding is very time consuming and costly, it was the only method available, till some years ago, to enhance the genetic characteristics of animals. For larger animals with long gestation period, it might take several decades to create a desired character by conventional breeding. With the advent of modern biotechnology, it is now possible to carry out manipulations at the genetic level to get the desired characteristics in animals. Transgenesis refers to the phenomenon of introduction of exogeneous

DNA into the genome to create and maintain a stable heritable character. The foreign DNA that is introduced is called transgene. And the animal whose genome is altered by adding one or more transgenes is said to be transgenic. The transgenes behave like other genes present in the animals genome and are passed on to the offsprings. Thus, transgenic animals are genetically engineered or genetically modified organisms (GMOs) with a new heritable character. It was in 1980s, the genetic manipulation of animals by introducing genes into fertilised eggs became a reality.

20.2   Importance of transgenic animals

Transgenesis has now become a powerful tool for studying the gene expression and developmental processes in higher organisms, besides the improvement in their genetic characteristics. Transgenic animals serve as good models for understanding the human diseases.

Further, several proteins produced by transgenic animals are important for medical and pharmaceutical applications. Thus, the transgenic farm animals are a part of the lucrative world-wide biotechnology industry, with great benefits to mankind. Transgenesis is important for improving the quality and quantity of milk, meat, eggs and wool production, besides creating drug resistant animals.

20.2.1   Milk as the medium of protein production

Milk is the secretion of mammary glands that can be collected frequently without causing any harm to the animal. Thus, milk from the transgenic animals can serve as a good and authenticated source of human proteins for a wide range of applications. Another advantage with milk is that it contains only a few proteins (casein, lactalbumin, immunoglobulin, etc.), in the native state, therefore isolation and purification of a new protein from milk is easy.

20.2.2   Commonly used animals for transgenesis

The first animals used for transgenesis was a mouse. The 'super mouse', was created by inserting a rat gene for growth hormone into the mouse genome. The offspring was much larger than the parents. Super mouse attracted a lot of public attention, since it was a product of genetic manipulation rather than the normal route of sexual reproduction. Mouse continues to be an animal of choice for most transgenic experiments. The other animals used for transgenesis include rat, rabbit, pig, cow, goat, sheep and fish.

Position effects

Position effect is the phenomenon of different levels of gene expression that is observed after insertion of a new gene at different position in the eukaryotic genome. This is commonly observed in transgenic animals as well as plants.

These transgenic organisms show variable levels and patterns of transgene expression. In a majority of cases, position effects are dependent on the site of transgene integration. In general, the defective expression is due to the insertion of transgene into a region of highly packed chromatin. The transgene will be more active if inserted into an area of open chromatin. The positional effects are overcome by a group of DNA sequences called insulators. The sequences referred to as specialised chromatin structure (SCS) are known to perform the functions of insulators. It has been demonstrated that the expression of the gene is appropriate if the transgene is flanked by insulators.

Animal bioreactors

Transgenesis is wonderfully utilised for production proteins of pharmaceutical and medical use. In fact, any protein synthesised in the human body can be made in the transgenic animals, provided that the genes are correctly programmed. The advantage with transgenic animals is to produce scarce human proteins in huge quantities.

Thus, the animals serving as factories for production of biologically important products are referred to as animal bioreactors or sometimes pharm animals. The transgenic animals as bioreactors can be commercially exploited for the benefit of mankind. Once developed, animal bioreactors are cost-effective for the production of large quantities of human proteins. Routine breeding and healthful living conditions are enough to maintain transgenic animals.

Transgenic animals in xenotransplantation

Organ transplantation (kidney, liver, heart, etc.), in humans has now become one of the advanced surgical practices to replace the defective, non-functional or severally damaged organs. The major limitation of transplantation is the shortage of organ donors. This often results in long waiting times and many unnecessary deaths of organ failure patients.

Xenotransplantation refers to the replacement of failed human organs by the functional animal organs. The major limitation of xenotransplantation is the phenomenon of hyper acute organ rejection due to host immune system.

20.3   Methods used for creation of transgenic animals

The five principal methods used for creation of transgenic animals are discussed below: (i) physical transfection, (ii) chemical transfection, (iii) retrovirus-mediated gene transfer, (iv) virus vector and (v) DNA packaged inside a bacterium.

20.3.1   Physical transfection

This method involves the direct microinjection of a chosen gene construct (a single gene or a combination of genes) from another member of the same species or from a different species, into the pro-nucleus of a fertilised ovum. It is one of the first methods that proved to be effective in mammals. The introduced DNA may lead to the over- or under-expression of certain genes or to the expression of genes entirely new to the animal species. The insertion of DNA is, however, a random process and there is a high probability that the introduced gene will not insert itself into a site on the host DNA that will permit its expression. The manipulated fertilised ovum is transferred into the oviduct of a recipient female or foster mother that has been induced to act as a recipient by mating with a vasectomised male. A major advantage of this method is its applicability to a wide variety of species. Other transfection methods include particle bombardment, ultrasound and electroporation.

20.3.2 Chemical transfection

There are several chemical transfection techniques for animal cells but all are based on similar principles. The calcium phosphate mediated DNA uptake involves the formation of a co-precipitate which is taken up by endocytosis. The ability of mammalian cells to take up exogenous DNA from the cultured medium was first reported in 1962.

Formation of a fine DNA and calcium phosphate co-precipitate (should be prepared fresh) facilitates DNA uptake by endocytosis. Some of the DNA fragments which enter the cell may reach the nucleus and integrated.

Expression of such genes confers the transfection. The transformation frequency of calcium precipitate method is generally low (1–2%), therefore use of soluble complexes (polyplexes) or liposomes and lipoplexes (fusogenic) phospholipid are used to package DNA inside these vehicles. The desired gene is transferred in plasmid and it may be used directly for chemical transfection or inserted in a bacterium for delivery into a mammalian cell. Yeast cells with the cell wall removed (spheroplast) therefore have been used to introduce YAC DNA into mouse, using liposomes and embryonic stem cells for the production of YAC transgenic mice.

20.3.3 Retrovirus-mediated gene transfer

To increase the probability of expression, gene transfer is mediated by means of a carrier or vector, generally a virus or a plasmid. Retroviruses are commonly used as vectors to transfer genetic material into the cell, taking advantage of their ability to infect host cells in this way. Offspring derived from this method are chimeric, i.e., not all cells carry the retrovirus. Transmission of the transgene is possible only if the retrovirus integrates into some of the germ cells and forms rPIC complex. For any of these techniques the success rate in terms of live birth of animals containing the transgene is extremely low. Providing that the genetic manipulation does not lead to abortion, the result is a first generation

(F1) of animals that need to be tested for the expression of the transgene. Depending on the technique used, the F1 generation may result in chimeras. When the transgene has integrated into the germ cells, the so called germ line chimeras are then inbred for 10 to 20 generations until homozygous transgenic animals are obtained and the transgene is present in every cell. At these stage embryos carrying the transgene can be frozen and stored for subsequent implantation.

There have been numerous reports and applications of transgenic plants in agriculture, mainly to benefit the producer. However, the realisation of genetically engineered livestock has been much slower. The production of transgenic animals has focused mainly on producing models (e.g., the mouse) for basic and medical research. In terms of commercially important livestock species, work has revolved around specialised non-agricultural purposes such as pharmaceutical production and xenotransplantation and to a lesser extent, applied agricultural purposes to improve animal production traits and animal-food products. In this case, one of the most important production animals, the dairy cow, was given enhanced resistance to a common and often devastating infection of the mammary gland, a potential benefit to both the producer and the animals well-being.

20.3.4 Virus vector

Viruses have a natural ability to absorb to the surface of host cells and infect. This property can be exploited to deliver rDNA into animal cells. The viral system is efficient in transfer, expressed well and replicate rapidly in the host cells. Several classes of viral vectors have been developed for use in human gene therapy and at least eight have been used in clinical trials.

General properties of viral vector are:

1. Transgene may be incorporated into viral vectors as additional gene or as replacement to certain genes of viral genome by ligation or homology recombination. If virus can propagate independently it is called helper independent. If essential viral genes are replaced by transgene then virus need a replication gene in trans position (another virus similar to binary vectors) and virus is called 'helper-dependent'.

2. It is necessary to prevent replication as well as recombination of helper virus. Icosahedral viruses such as adenovirus and retroviruses package their genome into preformed capsid, their volume is fixed with defined amount of DNA can be packaged. Rod shaped baculo-viruses form the capsid around the genome, so there are no such size constraints. There is no ideal virus, each has his own advantages or disadvantages.

Adenoviruses are DNA viruses with a linear, double stranded genome of approximately 36 kb. They are used in gene transfer because they show some advantageous features like stability, a high capacity for foreign DNA, a wide host range that includes non dividing cells and the ability to produce high titer stock (up to 10^{11} pfu/ml). They are suitable for transient expression in dividing cells because they do not integrate efficiently into the genome.

Adenoviruses are also used as gene therapy vectors because the virions are taken up efficiently by cells *in vivo* and adenovirus derived vaccines have been used in humans with no reported side effects. Baculo-viruses have large double stranded DNA genomes. They efficiently infect arthropods, particularly insects. Nuclear polyhedrosis viruses, a group of baculoviruses, have an unusual infection cycle that involves the production of nuclear occlusion bodies. These are pro-teinaceous particles in which the virions are embedded allowing the virus to survive harsh environmental conditions such as desiccation.

Baculo-viruses are mainly used for high level transient protein expression in insects and insect cells. Two baculoviruses have been extensively developed as vectors, namely the Autographa calofornica multiple nuclear polyhedrosis virus and the Bombyx mori nuclear polyhedrosis virus.

20.3.5   DNA packaged inside a bacterium (Bactofection)

Generally, *Agrobacterium* tumefaciens mediated transfer of DNA is a common practice in plant system. It has been shown by Kunik and co -workers that this bacterium can transfer DNA in cultured human cells. It was established in mid-1990 that several bacteria infect human cells and undergo lysis releasing plasmid in host cells, e.g., *Salmonella* sp., *Listeria* sp., and *Shigella* sp. The plasmid DNA then finds its way to the host cell nucleus, where it is integrated in the genome and expressed.

Another method of DNA transfer is by conjugation [the transfer of DNA through a pilus (plural pilli)]. This pilus is formed by bacterial cell. When live bacteria are used, it is necessary that the bacteria are attenuated. This is because the gene transfer system uses the natural ability of bacteria to infect eukaryotic cells. The bacteria may multiply and destroy host cells.

20.4   Application possibilities for gene transfer

In recent years, several application possibilities for gene transfer in domestic animals have been discussed. Until now it has been possible to influence only traits that are based on a single gene or on a limited number of genes. There are only a very limited number of traits of interest to breeders, which are based on a single gene. Various markers used to identify the transgenic animals are presented in Table 20.1. As far as gene transfer in cattle is concerned, there are realistic prospects that it will be possible to influence positively different production traits. Traditional selection programmes using conventional breeding techniques have achieved important results and will continue to do so. However, it seems preferable to concentrate the very expensive and complex technique of gene transfer to fields which until now could only be improved with limited success through conventional breeding programmes, such as breeding for increased disease resistance.

Table 20.1: Various markers used in transgenic animals.

Marker	Product	Selection
Ada	Adenosine deaminase	Xyl-A (9-β-D-xylofurenesyl adenosine) and 2′-deoxycoformycin
Tk	Thyrnidine kinase	Thynidine and aminopterin to block *de novo* dTTP synthesis selected on HAT medium
Ble	Glycopeptide binding progein	Confers resistance to glycopeptides. antibiotics bleomycin. plcomycin, pheomycin, Zeocin
His D	Histidinol dehydrogenase	Confers resistance to histidinol
Hpt	Hygromyein	Confers resistance to hygromycin
Npt II	Neomycin phosphotransftrase	Confers resistance to kanamycin, neomycin
Pac	Puromycin *N*-acetyl transferase	Confers resistance to puromycin
trpB	Tryptophan synthase	Confers resistance to indole

20.5   Benefits of transgenic animals

The benefits of transgenic animals are discussed below: (i) normal physiology and development, (ii) study of disease, (iii) biological products, (iv) vaccine safety and (v) chemical safety testing.

20.5.1   Normal physiology and development

Transgenic animals can be specifically designed to allow the study of how genes are regulated and how they affect the normal functioning of the body and its development. For example, the study of complex factors involved in growth such as insulin likes growth factors.

20.5.2   Study of disease

Many transgenic animals are designed to increase the understanding of that how genes contribute to the development of diseases such as cancer, cystic fibrosis, rheumatoid arthritis and Alzheimer. These are specially made to serve as models for human diseases, so that investigation of new treatments for diseases is made possible.

20.5.3   Biological products

Human disease can be treated by medicines that contain biological products.

1. Transgenic animals that produce useful biological products can be created by the introduction of the portion of the DNA or genes that codes for a

particular product such as human protein (alpha-1-antitrypsin) which is used to treat emphysema.

2. Similar attempts are being made for the treatment of phenylketonuria (PKU) and cystic fibrosis. For example, the first transgenic cow Rosie produced human protein enriched milk (2.4 g/L) in 1997. The milk contained the human alpha lactalbumin and was nutritionally a more balanced product for human babies than natural cow milk.

20.5.4 Vaccine safety

Transgenic mice are being used in testing the safety of vaccines before they are used in humans, e.g., polio vaccine. These animals are also used for the toxicity or safety testing procedures. If found reliable and successful they could replace the use of monkeys in order to test the safety of batches of the vaccine.

20.5.5 Chemical safety testing

Transgenic animals are made to carry the genes, which makes them more sensitive to the toxic substances than the non-transgenic ones. They are then exposed to toxic substances and effects are studied. This is known as toxicity/safety testing.

20.6 Genetically modified mouse

A genetically modified mouse is a mouse that has had its genome altered through the use of genetic engineering techniques. Genetically modified mice are commonly used for research or as animal models of human diseases.

20.6.1 Methods to produce genetically modified mice

There are two basic technical approaches to produce genetically modified mice. The first involves pronuclear injection into a single cell of the mouse embryo, where it will randomly integrate into the mouse genome. This method creates a transgenic mouse and is used to insert new genetic information into the mouse genome or to over-express endogenous genes. The second approach, pioneered by Oliver Smithies and Mario Capecchi, involves modifying embryonic stem cells with a DNA construct containing DNA sequences homologous to the target gene. Embryonic stem cells that recombine with the genomic DNA are selected for and they are then injected into the mice blastocysts. This method is used to manipulate a single gene, in most cases 'knocking out' the target gene, although more subtle genetic manipulation can occur (e.g., only changing single nucleotides).

20.6.2 Uses of genetically modified mice

Genetically modified mice are used extensively in research as models of human disease. The most common type is the knockout mouse, where the activity of a single (or in some cases multiple) genes are removed. They have been used to study and model obesity, heart disease, diabetes, arthritis, substance abuse, anxiety, ageing and Parkinson disease. Transgenic mice (Fig. 20.1) generated to carry cloned oncogenes and knockout mice lacking tumour suppressing genes have provided good models for human cancer. Hundreds of these oncomice have been developed covering a wide range of cancers affecting most organs of the body and they are being refined to become more representative of human cancer. The disease symptoms and potential drugs or treatments can be tested against these mouse models.

Figure 20.1: Transgenic mice.

A mouse has been genetically engineered to have increased muscle growth and strength by overexpressing the insulin-like growth factor I (IGF-I) in differentiated muscle fibres. Another mouse has had a gene altered that is involved in glucose metabolism and runs faster, lives longer, is more sexually active and eats more without getting fat than the average mouse.

Great care should be taken when deciding how to use genetically modified mice in research. Even basic issues like choosing the correct 'wild-type' control mouse to use for comparison are sometimes overlooked.

20.7 Contribution of transgenic animals to human welfare

The benefits of these animals to human welfare can be grouped into areas:

1. Agriculture.
2. Medicine.
3. Industry.

20.7.1 Agricultural applications of transgenic animals

Breeding

Farmers have always used selective breeding to produce animals that exhibit desired traits (e.g., increased milk production, high growth rate). Traditional breeding is a time-consuming, difficult task. When technology using molecular biology was developed, it became possible to develop traits in animals in a shorter time and with more precision. In addition, it offers the farmer an easy way to increase yields.

Quality

Transgenic cows exist that produce more milk or milk with less lactose or cholestero, pigs and cattle that have more meat on them and sheep that grow more wool. In the past, farmers used growth hormones to spur the development of animals but this technique was problematic, especially since residue of the hormones remained in the animal product.

Disease resistance

Scientists are attempting to produce disease-resistant animals, such as influenza-resistant pigs, but a very limited number of genes are currently known to be responsible for resistance to diseases in farm animals.

20.7.2 Medical applications of transgenic animals

Xenotransplantation

Patients die every year for lack of a replacement heart, liver, or kidney. For example, about 5000 organs are needed each year in the United Kingdom alone. Transgenic pigs may provide the transplant organs needed to alleviate the shortfall. Currently, xenotransplantation is hampered by a pig protein that can cause donor rejection but research is underway to remove the pig protein and replace it with a human protein.

Nutritional supplements and pharmaceuticals

Products such as insulin, growth hormone and blood anti-clotting factors may soon be or have already been obtained from the milk of transgenic cows, sheep, or goats. Research is also underway to manufacture milk through transgenesis for treatment of debilitating diseases such as phenylketonuria (PKU), hereditary emphysema and cystic fibrosis. In 1997, the first transgenic cow, Rosie, produced human protein-enriched milk at 2.4 grams per litre. This transgenic milk is a more nutritionally balanced product than natural bovine milk and could be given to babies or the elderly with special nutritional or digestive needs. Rosie's milk contains the human gene alpha-lactalbumin.

Human gene therapy

Human gene therapy involves adding a normal copy of a gene (transgene) to the genome of a person carrying defective copies of the gene. The potential for treatments for the 5000 named genetic diseases is huge and transgenic animals could play a role. For example, the A. I. Virtanen Institute in Finland produced a calf with a gene that makes the substance that promotes the growth of red cells in humans.

20.7.3   Ethical concerns surrounding transgenesis

This section focuses on the benefits of the technology; however, thoughtful ethical decision-making cannot be ignored by the biotechnology industry, scientists, policy-makers and the public.

These ethical issues, include questions such as:

1. Should there be universal protocols for transgenesis?
2. Should such protocols demand that only the most promising research be permitted?
3. Is human welfare the only consideration? What about the welfare of other life forms?
4. Should scientists focus on *in vitro* (cultured in a lab) transgenic methods rather than, or before, using live animals to alleviate animal suffering?
5. Will transgenic animals radically change the direction of evolution, which may result in drastic consequences for nature and humans alike?
6. Should patents be allowed on transgenic animals, which may hamper the free exchange of scientific research?

20.8   Genetically modified fish

Genetically modified fish (GM fish) are genetically modified organisms. The DNA of the fish has been modified using genetic engineering techniques. In most cases the aim is to introduce a new trait to the fish which does not occur naturally in the species. GM fish are used in scientific research and while they are being developed for use in aquaculture food production, as of May 2012 no GM fish has been approved by the FDA for this purpose. Some GM fish that have been created have promoters driving an over-production of 'all fish' growth hormone.

This resulted in dramatic growth enhancement in several species, including salmonids, carps and tilapias. Critics have objected to GM fish *per se* on several grounds, including ecological concerns and with respect to whether using them as food is safe and whether GM fish are needed to address the world's food needs.

20.8.1   Types of genetically modified fish

Salmon fish

Salmon belong to the Salmonidae family which also includes salmon and trout. Although the smallest species is just 13 centimeters (5.1 in) long as an adult, most are much larger and the largest can reach 2 meters (6.6 ft). All salmonids spawn in fresh water, but they spend most of their maturity in the sea. This life style is known as anadromous.

They are considered to be predators, because they feed on small crustaceans, aquatic insects and smaller fish. A genetically modified Atlantic salmon known as the AquAdvantage salmon has an increased growth rate and size over the wild type Atlantic salmon from which it was derived, up to doubling its weight with a reduced time of growth to maturity.

Tilapia fish

Tilapia is the common name for several species of cichlid fish from the tilapine cichlid tribe. Tilapia inhabit a wide range fresh water habitats, including lakes, streams, ponds and rivers. Anciently, tilapia hold great significance in artisan fishing in Africa and are paramount lately in aquaculture. Tilapia are very vulnerable to cold temperatures and thus survive well with temperatures above 60°F (16°C). Tilapia is the fifth most important fish in fish farming. Because of their large size, rapid growth and palatability, tilapiine cichlids are the focus of major farming efforts, specifically various species.

Zebrafish

Zebrafish are freshwater fish and are part of the Cyprinidae. They are a popular aquarium fish, commonly sold as zebra danio and have been very vital as model organisms in research. They derive their name from the uniform horizontal stripes along the side of the body bilaterally. Males bear gold stripes within the blue stripes, while females bear silver stripes within the blue stripes. Zebrafish can mature up to 6.4 centimeters in the wild, but usually it is rare for them to mature beyond 4″ in captivity.

20.8.2   Uses of genetically modified fish

Most genetically engineered fish are used in basic research in genetics and development. Two species of fish, zebrafish and medaka, are most commonly modified because they have optically clear chorions (shells), rapidly develop and the 1-cell embryo is easy to see and microinject with transgenic DNA. They are also used in drug discovery. Also, zebrafish have the capability of regenerating their organ tissues and GM zebrafish are being explored for benefits of unlocking human organ tissue diseases and failure mysteries. For instance zebrafish are used to understand heart tissue repair and regeneration in efforts to study and discover cures for cardiovascular diseases.

Pets fish

The GloFish is a patented brand of genetically modified (GM) fluorescent zebrafish with bright red, green and orange fluorescent colour. Although not originally developed for the ornamental fish trade, it became the first genetically modified animal to become publicly available as a pet when it was introduced for sale in 2003. They were quickly banned for sale in California.

Food (potential)

Genetically modified fish have been developed with promoters driving an over-production of 'all fish' growth hormone for use in the aquaculture industry to increase the speed of development and potentially reduce fishing pressure on wild stocks. This has resulted in dramatic growth enhancement in several species, including salmon, trout and tilapia.

AquaBounty, the leading company in GM fish for the food industry, claims that their GM AquAdvantage Salmon can mature in half the time it takes non-GM salmon and achieves twice the size.

Detecting aquatic pollution (potential)

Several academic groups have been developing GM zebrafish to detect aquatic pollution. The lab that originated the GloFish discussed above originally developed them to change colour in the presence of pollutants, to be used as environmental sensors. A lab at University of Cincinnati has been developing GM zebrafish for the same purpose, as has a lab at Tulane University.

20.8.3 Regulation of genetic engineering

The regulation of genetic engineering concerns the approaches taken by governments to assess and manage the risks associated with the development and release of genetically modified crops. There are differences in the regulation of GMOs between countries, with some of the most marked differences occurring between the U.S. and Europe. Regulation varies in a given country depending on the intended use of the products of the genetic engineering. For example, a fish not intended for food use is generally not reviewed by authorities responsible for food safety.

20.8.4 Controversy of genetic engineering

Critics have objected to use of genetic engineering *per se* on several grounds, including ethical concerns, ecological concerns (especially about gene flow) and economic concerns raised by the fact GM techniques and GM organisms are subject to intellectual property law. GMOs also are involved in controversies over GM food with respect to whether using GM fish as safe is safe, whether it would exacerbate or cause fish allergies, whether it should be labelled and whether GM fish and crops are needed to address the world's food needs. These controversies have led to litigation, international trade disputes and protests and to restrictive regulation of commercial products in most countries. With respect to concerns about GM AquAdvantage Salmon interbreeding with wild fish, the company indicates that their GM salmon are not capable of reproducing, as the GM fish are sterile. AquaBounty also emphasises that their GM fish would not survive wild conditions due to the geographical locations where their research is being done, as well as the locations of their farms.

20.9   Stem cell

Stem cells are undifferentiated biological cells that can differentiate into specialised cells and can divide (through mitosis) to produce more stem cells. They are found in multicellular organisms. In mammals, there are two broad types of stem cells: embryonic stem cells, which are isolated from the inner cell mass of blastocysts and adult stem cells, which are found in various tissues. In adult organisms, stem cells and progenitor cells act as a repair system for the body, replenishing adult tissues. In a developing embryo, stem cells can differentiate into all the specialised cells — ectoderm, endoderm and mesoderm — but also maintain the normal turnover of regenerative organs, such as blood, skin, or intestinal tissues.

There are three known accessible sources of autologous adult stem cells in humans:

1. Bone marrow, which requires extraction by harvesting, that is, drilling into bone (typically the femur or iliac crest).

2. Adipose tissue (lipid cells), which requires extraction by liposuction.

3. Blood, which requires extraction through apheresis, wherein blood is drawn from the donor (similar to a blood donation) and passed through a machine that extracts the stem cells and returns other portions of the blood to the donor.

Stem cells can also be taken from umbilical cord blood just after birth. Of all stem cell types, autologous harvesting involves the least risk. By definition, autologous cells are obtained from one's own body, just as one may bank his or her own blood for elective surgical procedures. Adult stem cells are frequently used in medical therapies, for example in bone marrow transplantation. Stem cells can now be artificially grown and transformed (differentiated) into specialised cell types with characteristics consistent with cells of various tissues such as muscles or nerves. Embryonic cell lines and autologous embryonic stem cells generated through Somatic-cell nuclear transfer or dedifferentiation have also been proposed as promising candidates for future therapies.

20.9.1 Properties of stem cell

The classical definition of a stem cell requires that it possess two properties:

1. Self-renewal: The ability to go through numerous cycles of cell division while maintaining the undifferentiated state.
2. Potency: The capacity to differentiate into specialised cell types. In the strictest sense, this requires stem cells to be either totipotent or pluripotent — to be able to give rise to any mature cell type, although multipotent or unipotent progenitor cells are sometimes referred to as stem cells. Apart from this it is said that stem cell function is regulated in a feedback mechanism.

Self-renewal

Two mechanisms exist to ensure that a stem cell population is maintained:

1. Obligatory asymmetric replication: A stem cell divides into one mother cell that is identical to the original stem cell and another daughter cell that is differentiated.
2. Stochastic differentiation: When one stem cell develops into two differentiated daughter cells, another stem cell undergoes mitosis and produces two stem cells identical to the original.

Potency definition

Potency specifies the differentiation potential (the potential to differentiate into different cell types) of the stem cell.

Identification

In practice, stem cells are identified by whether they can regenerate tissue. For example, the defining test for bone marrow or hematopoietic stem cells (HSCs) is the ability to transplant the cells and save an individual without HSCs. This demonstrates that the cells can produce new blood cells over a long term. It should also be possible to isolate stem cells from the transplanted individual, which can themselves be transplanted into another individual without HSCs, demonstrating that the stem cell was able to self-renew. Properties of stem cells can be illustrated *in vitro*, using methods such as clonogenic assays, in which single cells are assessed for their ability to differentiate and self-renew.

Stem cells can also be isolated by their possession of a distinctive set of cell surface markers. However, *in vitro* culture conditions can alter the behaviour of cells, making it unclear whether the cells will behave in a similar manner *in vivo*. There is considerable debate as to whether some proposed adult cell populations are truly stem cells.

20.9.2 Embryonic stem (ES) cells

Embryonic stem (ES) cells are stem cells derived from the inner cell mass of a blastocyst, an early-stage embryo. Human embryos reach the blastocyst stage 4–5 days post fertilisation, at which time they consist of 50–150 cells. ES cells are pluripotent and give rise during development to all derivatives of the three primary germ layers: ectoderm, endoderm and mesoderm. In other words, they can develop into each of the more than 200 cell types of the adult body when given sufficient and necessary stimulation for a specific cell type. They do not contribute to the extra-embryonic membranes or the placenta.

Nearly all research to date has made use of mouse embryonic stem cells (mES) or human embryonic stem cells (hES). Both have the essential stem cell characteristics, yet they require very different environments in order to maintain an undifferentiated state. Mouse ES cells are grown on a layer of gelatin as an extracellular matrix (for support) and require the presence of leukemia inhibitory factor (LIF). Human ES cells are grown on a feeder layer of mouse embryonic fibroblasts (MEFs) and require the presence of basic fibroblast growth factor (bFGF or FGF-2). Without optimal culture conditions or genetic manipulation, embryonic stem cells will rapidly differentiate.

A human embryonic stem cell is also defined by the expression of several transcription factors and cell surface proteins. The transcription factors Oct-4, Nanog and Sox2 form the core regulatory network that ensures the suppression of genes that lead to differentiation and the maintenance of pluripotency. The cell surface antigens most commonly used to identify hES cells are the glycolipids stage specific embryonic antigen 3 and 4 and the keratan sulphate antigens Tra-1-60 and Tra-1-81. The molecular definition of a stem cell includes many more proteins and continues to be a topic of research.

Fetal stem cells

The primitive stem cells located in the organs of fetuses are referred to as fetal stem cells. There are two types of fetal stem cells:

1. Fetal proper stem cells come from the tissue of the fetus proper and are generally obtained after an abortion. These stem cells are not immortal but have a high level of division and are multipotent.

2. Extraembryonic fetal stem cells come from extraembryonic membranes and are generally not distinguished from adult stem cells. These stem cells are acquired after birth, they are not immortal but have a high level of cell division and are pluripotent.

Adult stem cells

Adult stem cells, also called somatic stem cells, are stem cells which maintain and repair the tissue in which they are found. They can be found in children, as well as adults. Pluripotent adult stem cells are rare and generally small in number, but they can be found in umbilical cord blood and other tissues. Bone marrow is a rich source of adult stem cells, which have been used in treating several conditions including spinal cord injury, liver cirrhosis, chronic limb ischemia and endstage heart failure. The quantity of bone marrow stem cells declines with age and is greater in males than females during reproductive years. Much adult stem cell research to date has aimed to characterise their potency and self-renewal capabilities. In mice, pluripotent stem cells are directly generated from adult fibroblast cultures. However, mice do not live long with stem cell organs.

Most adult stem cells are lineage-restricted (multipotent) and are generally referred to by their tissue origin (mesenchymal stem cell, adipose-derived stem cell, endothelial stem cell, dental pulp stem cell, etc.). Stem cell division and differentiation is shown in Fig. 20.3. Adult stem cell treatments have been successfully used for many years to treat leukemia and related bone/blood cancers through bone marrow transplants.

Adult stem cells are also used in veterinary medicine to treat tendon and ligament injuries in horses.

The use of adult stem cells in research and therapy is not as controversial as the use of embryonic stem cells, because the production of adult stem cells does not require the destruction of an embryo. Additionally, in instances where adult stem cells are obtained from the intended recipient (an autograft), the risk of rejection is essentially non-existent. Consequently, more U.S. government funding is being provided for adult stem cell research.

Amniotic stem cells

Multipotent stem cells are also found in amniotic fluid. These stem cells are very active, expand extensively without feeders and are not tumorigenic. Amniotic stem cells are multipotent and can differentiate in cells of adipogenic, osteogenic, myogenic, endothelial, hepatic and also neuronal lines. Amniotic stem cells are a topic of active research. Use of stem cells from amniotic fluid overcomes the ethical objections to using human embryos as a source of cells. Roman Catholic teaching forbids the use of embryonic stem cells in experimentation.

Cord blood stem cells

A certain kind of cord blood stem cell (CB-SC) is multipotent and displays embryonic and hematopoietic characteristics. Phenotypic characterisation

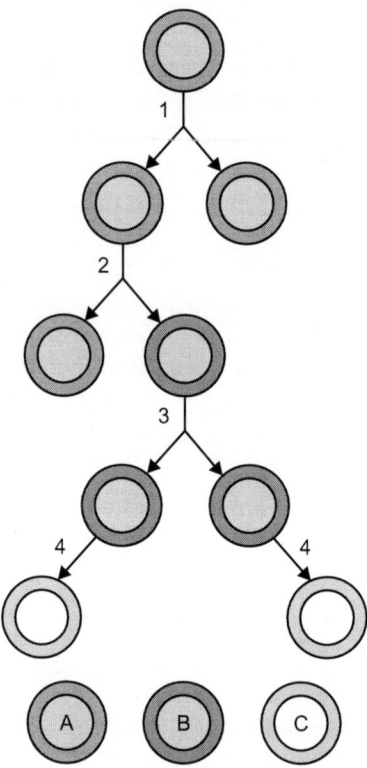

Figure 20.2: Stem cell division and differentiation. (A) stem cell, (B) progenitor cell, (C) differentiated cell, 1. symmetric stem cell division, 2. asymmetric stem cell division, 3. progenitor division, 4. terminal differentiation.

demonstrates that (CB-SCs) display embryonic cell markers (e.g., transcription factors OCT-4 and Nanog, stage-specific embryonic antigen (SSEA)-3 and SSEA-4) and leukocyte common antigen CD45, but that they are negative for blood cell lineage markers (e.g., CD1a, CD3, CD4, CD8, CD11b, CD11c, CD13, CD14, CD19, CD20, CD34, CD41a, CD41b, CD83, CD90, CD105 and CD133).

Additionally, CB-SCs display very low immunogenicity as indicated by expression of a very low level of major histocompatibility complex (MHC) antigens and failure to stimulate the proliferation of allogeneic lymphocytes. They can give rise to three embryonic layer-derived cells in the presence of different inducers. More specifically, CB-SCs tightly adhere to culture dishes with a large rounded morphology and are resistant to common detaching methods (trypsin/EDTA). CB-SCs are the active agent in stem cell educator therapy, which has therapeutic potential against autoimmune diseases like type 1 diabetes according to studies by Yong Zhao and others.

Induced pluripotent stem cells

These are not adult stem cells, but rather adult cells (e.g., epithelial cells) reprogrammed to give rise to pluripotent capabilities. Using genetic reprogramming with protein transcription factors, pluripotent stem cells equivalent to embryonic stem cells have been derived from human adult skin tissue. Shinya Yamanaka and his colleagues at Kyoto University used the transcription factors Oct3/4, Sox2, c-Myc and Klf4 in their experiments on cells from human faces. Junying Yu, James Thomson and their colleagues at the University of Wisconsin–Madison used a different set of factors, Oct4, Sox2, Nanog and Lin28 and carried out their experiments using cells from human foreskin. As a result of the success of these experiments, Ian Wilmut, who helped create the first cloned animal Dolly the Sheep, has announced that he will abandon somatic cell nuclear transfer as an avenue of research. Frozen blood samples can be used as a source of induced pluripotent stem cells, opening a new avenue for obtaining the valued cells.

Lineage stem cells

To ensure self-renewal, stem cells undergo two types of cell division (see Stem cell division and differentiation Fig. 20.2). Symmetric division gives rise to two identical daughter cells both endowed with stem cell properties. Asymmetric division, on the other hand, produces only one stem cell and a progenitor cell with limited self-renewal potential. Progenitors can go through several rounds of cell division before terminally differentiating into a mature cell. It is possible that the molecular distinction between symmetric and asymmetric divisions lies in differential segregation of cell membrane proteins (such as receptors) between the daughter cells.

An alternative theory is that stem cells remain undifferentiated due to environmental cues in their particular niche. Stem cells differentiate when they leave that niche or no longer receive those signals. Studies in *Drosophila germarium* have identified the signals decapentaplegic and adherens junctions that prevent germarium stem cells from differentiating.

20.9.3   *Cre-Lox* recombination

In the field of genetics, *Cre-Lox* recombination is known as a site-specific recombinase technology and is widely used to carry out deletions, insertions, translocations and inversions at specific sites in the DNA of cells. It allows the DNA modification to be targeted to a specific cell type or be triggered by a specific external stimulus. It is implemented both in eukaryotic and prokaryotic systems. The system consists of a single enzyme, *Cre* recombinase, that recombines a pair of short target sequences called the *Lox* sequences. This system can be implemented without inserting any extra supporting proteins or sequences.

The *Cre* enzyme and the original *Lox* site called the *LoxP* sequence are derived from bacteriophage P1.

Placing *Lox* sequences appropriately allows genes to be activated, repressed, or exchanged for other genes. At a DNA level many types of manipulations can be carried out. The activity of the *Cre* enzyme can be controlled so that it is expressed in a particular cell type or triggered by an external stimulus like a chemical signal or a heat shock. These targeted DNA changes are useful in cell lineage tracing and when mutants are lethal if expressed globally.

The *Cre-Lox* system is very similar in action and in usage to the FLP-FRT recombination system.

Molecular markers

21.1   Introduction

During the last few decades, the use of molecular markers, revealing polymorphism at the DNA level, has been playing an increasing part in plant biotechnology and their genetics studies. There are different types of markers, viz., morphological, biochemical and DNA based molecular markers. These DNA based markers are differentiates in two types first non PCR based (RFLP) and second is PCR based markers (RAPD, AFLP, SSR, SNP, etc.), amongst others, the microsatellite DNA marker has been the most widely used, due to its easy use by simple PCR, followed by a denaturing gel electrophoresis for allele size determination and to the high degree of information provided by its large number of alleles per locus. Despite this, a new marker type, named SNP, for Single Nucleotide Polymorphism, is now on the scene and has gained high popularity, even though it is only a bi-allelic type of marker. Day by day development of such new and specific types of markers makes their importance in understanding the genomic variability and the diversity between the same as well as different species of the plants. This chapter discusses the biochemical and molecular markers their advantages, disadvantages and the applications of the marker in comparison with other markers types.

In current scenario, the DNA markers become the marker of choice for the study of crop genetic diversity has become routine, to revolutionised the plant biotechnology. Increasingly, techniques are being developed to more precisely, quickly and cheaply assess genetic variation. In this reviews basic qualities of molecular markers, their characteristics, the advantages and disadvantages of their applications and analytical techniques and provides some examples of their use. There is no single molecular approach for many of the problems facing gene bank managers and many techniques complement each other. However, some techniques are clearly more appropriate than others for some specific applications like wise crop diversity and taxonomy studies.

Due to the rapid developments in the field of molecular genetics, varieties of different techniques have emerged to analyse genetic variation during the last few decayed. These genetic markers may differ with respect to important features, such as genomic abundance, level of polymorphism detected, locus specificity, reproducibility, technical requirements and financial investment. No marker is superior to all others for a wide range of applications. The most appropriate genetic marker has depend on the specific application, the presumed level of polymorphism, the presence of sufficient technical facilities and know how, time constraints and financial limitations.

21.2 Biochemical marker - allozymes (isozyme)

Isozymes analysis has been used for over 70 years for various research purposes in biology, viz., to delineate phylogenetic relationships, to estimate genetic variability and taxonomy, to study population genetics and developmental biology, to characterisation in plant genetic resources management and plant breeding. Isozymes are defined as structurally different molecular forms of an enzyme with, qualitatively, the same catalytic function. Isozymes originate through amino acid alterations, which cause changes in net charge, or the spatial structure (conformation) of the enzyme molecules and also, therefore, their electrophoretic mobility. After specific staining the isozyme profile of individual samples can be observed.

Allozymes are allelic variants of enzymes encoded by structural genes. Enzymes are proteins consisting of amino acids, some of which are electrically charged. As a result, enzymes have a net electric charge, depending on the stretch of amino acids comprising the protein. When a mutation in the DNA results in an amino acid being replaced, the net electric charge of the protein may be modified and the overall shape (conformation) of the molecule can change. Because of changes in electric charge and conformation can affect the migration rate of proteins in an electric field, allelic variation can be detected by gel electrophoresis and subsequent enzyme-specific stains that contain substrate for the enzyme, cofactors and an oxidised salt (e.g., nitro-blue tetrazolium). Usually two, or sometimes even more loci can be distinguished for an enzyme and these are termed isoloci. Therefore, allozyme variation is often also referred to as isozyme variation isozymes have been proven to be reliable genetic markers in breeding and genetic studies of plant species, due to their consistency in their expression, irrespective of environmental factors.

21.2.1 Advantages and disadvantages

Advantages

The strength of allozymes is simplicity. Because allozyme analysis does not require DNA extraction or the availability of sequence information, primers or probes, they are quick and easy to use. Some species, however, can require considerable optimisation of techniques for certain enzymes. Simple analytical procedures, allow some allozymes to be applied at relatively low costs, depending on the enzyme staining reagents used. Isoenzyme markers are the oldest among the molecular markers. Isozymes markers have been successfully used in several crop improvement programmes. Allozymes are codominant markers that have high reproducibility. Zymograms (the banding pattern of isozymes) can be readily interpreted in terms of loci and alleles, or they may require segregation analysis of progeny of known parental crosses for interpretation. Sometimes, however, zymograms present complex banding profiles arising from polyploidy or duplicated genes and the formation of intergenic heterodimers, which may complicate interpretation.

Disadvantages

The main weakness of allozymes is their relatively low abundance and low level of polymorphism. Moreover, proteins with identical electrophoretic mobility (co-migration) may not be homologous for distantly related germplasm. In addition, their selective neutrality may be in question. Lastly, often allozymes are considered molecular markers since they represent enzyme variants and enzymes are molecules. However, allozymes are in fact phenotypic markers and as such they may be affected by environmental conditions. For example, the banding profile obtained for a particular allozyme marker may change depending on the type of tissue used for the analysis (e.g., root vs. leaf). This is because a gene that is being expressed in one tissue might not be expressed in other tissues. On the contrary, molecular markers, because they are based on differences in the DNA sequence, are not environmentally influenced, which means that the same banding profiles can be expected at all times for the same genotype.

21.2.2   Applications of allozymes

Allozymes have been applied in many population genetics studies, including measurements of out crossing rates, (sub) population structure and population divergence. Allozymes are particularly useful at the level of conspecific populations and closely related species and are therefore useful to study diversity in crops and their relatives. They have been used, often in concert with other markers, for fingerprinting purposes and diversity studies, to study interspecific relationships, the mode of genetic inheritance and allelic frequencies in germplasm collections over serial increase cycles in germplasm banks and to identify parents in hybrids.

21.3   Molecular markers

A molecular markers a DNA sequence that is readily detected and whose inheritance can be easily be monitored. The uses of molecular markers are based on the naturally occurring DNA polymorphism, which forms basis for designing strategies to exploit for applied purposes. A marker must to be polymorphic, i.e., it must exit in different forms so that chromosome carrying the mutant genes can be distinguished from the chromosomes with the normal gene by a marker it also carries. Genetic polymorphism is defined as the simultaneous occurrence of a trait in the same population of two discontinuous variants or genotypes. DNA markers seem to be the best candidates for efficient evaluation and selection of plant material.

Unlike protein markers, DNA markers segregate as single genes and they are not affected by the environment. DNA is easily extracted from plant materials and its analysis can be cost and labour effective. The first such DNA markers to be utilised were fragments produced by restriction digestion–the restriction fragment length polymorphism (RFLP) based genes marker. Consequently, several markers system has been developed.

An ideal molecular marker must have some desirable properties.

1. Highly polymorphic nature: It must be polymorphic as it is polymorphism that is measured for genetic diversity studies.
2. Codominant inheritance: Determination of homozygous and heterozygous states of diploid organisms.
3. Frequent occurrence in genome: A marker should be evenly and frequently distributed throughout the genome.
4. Selective neutral behaviours: The DNA sequences of any organism are neutral to environmental conditions or management practices.
5. Easy access (availability): It should be easy, fast and cheap to detect.
6. Easy and fast assay.
7. High reproducibility.
8. Easy exchange of data between laboratories.

It is extremely difficult to find a molecular marker, which would meet all the above criteria. A wide range of molecular techniques is available that detects polymorphism at the DNA level. Depending on the type of study to be undertaken, a marker system can be identified that would fulfill at least a few of the above characteristics. Various types of molecular markers are utilised to evaluate DNA polymorphism and are generally classified as hybridisation-based markers and polymerase chain reaction (PCR)-based markers. In the former, DNA profiles are visualised by hybridising the restriction enzyme-digested DNA, to a labelled probe, which is a DNA fragment of known origin or sequence. PCR based markers involve *in vitro* amplification of particular DNA sequences or loci, with the help of specifically or arbitrarily chosen oligonucleotide sequences (primers) and a thermos table DNA polymerase enzyme. The amplified fragments are separated electrophoretically and banding patterns are detected by different methods such as staining and autoradiography. PCR is a versatile technique invented during the mid-1980s. Ever since thermos table DNA polymerase was introduced in 1988, the use of PCR in research and clinical laboratories has increased tremendously. The primer sequences are chosen to allow base-specific binding to the template in reverse orientation. PCR is extremely sensitive and operates at a very high speed. Its application for diverse purposes has opened up a multitude of new possibilities in the field of molecular biology.

21.4   Restriction fragment length polymorphism

Restriction Fragment Length Polymorphism (RFLP) is a technique in which organisms may be differentiated by analysis of patterns derived from cleavage of their DNA. If two organisms differ in the distance between sites of cleavage of particular *Restriction Endonucleases*, the length of the fragments produced will differ when the DNA is digested with a restriction enzyme. The similarity of the patterns generated can be used to differentiate species (and even strains) from one another. This technique is mainly based on the special class of enzyme, i.e., Restriction Endonucleases. They have their origin in the DNA rearrangements that occur due to evolutionary processes, point mutations within the restriction enzyme recognition site sequences, insertions or deletions within the fragments and unequal crossing over. Size fractionation is achieved by gel electrophoresis and, after transfer to a membrane by Southern blotting; fragments of interest are identified by hybridisation with radioactive labelled probe. Different sizes or lengths of restriction fragments are typically produced when different individuals are tested. Such a polymorphism can by used to distinguish plant species, genotypes and, in some cases, individual plants. In RFLP analysis, restriction enzyme-digested genomic DNA is resolved by gel electrophoresis and then blotted on to a nitrocellulose membrane. Specific banding patterns are then visualised by hybridisation with labelled probe. Labelling of the probe may be performed with a radioactive isotope or with alternative non-radioactive stains, such as digoxigenin or fluorescein. These probes are mostly species-specific single locus probes of about 0.5–3.0 kb in size, obtained from a cDNA library or a genomic library. Though genomic library probes may exhibit greater variability than gene probes from cDNA libraries, a few studies reveal the converse.

21.4.1   Advantages and disadvantages

Advantages

RFLPs are generally found to be moderately polymorphic. In addition to their high genomic abundance and their random distribution, RFLPs have the advantages of showing codominant alleles and having high reproducibility. RFLP markers were used for the first time in the construction of genetic maps by Botstein and others. RFLPs, being codominant markers, can detect coupling phase of DNA molecules, as DNA fragments from all homologous chromosomes are detected. They are very reliable markers in linkage analysis and breeding and can easily determine if a linked trait is present in a homozygous or heterozygous state in individual, information highly desirable for recessive traits.

Disadvantages

The of utility RFLPs has been hampered due to the large quantities $(1-10\,\mu g)$ of purified, high molecular weight DNA are required for each DNA digestion and Southern blotting. Larger quantities are needed for species with larger genomes and for the greater number of times needed to probe each blot. The requirement of radioactive isotope makes the analysis relatively expensive and hazardous. The assay is time-consuming and labourintensive and only one out of several markers may be polymorphic, which is highly inconvenient especially for crosses between closely related species. Their inability to detect single base changes restricts their use in detecting point mutations occurring within the regions at which they are detecting polymorphism.

21.4.2   Applications

RFLPs can be applied in diversity and phylogenetic studies ranging from individuals within populations or species, to closely related species. RFLPs have been widely used in gene mapping studies because of their high genomic abundance due to the ample availability of different restriction enzymes and random distribution throughout the genome. They also have been used to investigate relationships of closely related taxa, as fingerprinting tools, for diversity studies and for studies of hybridisation and introgression, including studies of gene flow between crops and weeds. RFLP markers were used for the first time in the construction of genetic maps by Botstein and others. A set of RFLP genetic markers provided the opportunity to develop a detailed genetic map of lettuce.

21.5   Random amplified polymorphic DNA (RAPD)

RAPD is a PCR-based technology. The method is based on enzymatic amplification of target or random DNA segments with arbitrary primers. In 1991 Welsh and McClelland developed a new PCR-based genetic assay namely randomly amplified polymorphic DNA (RAPD). This procedure detects nucleotide sequence polymorphisms in DNA by using a single primer of arbitrary nucleotide sequence. In this reaction, a single species of primer anneals to the genomic DNA at two different sites on complementary strands of DNA template. If these priming sites are within an amplifiable range of each other, a discrete DNA product is formed through thermo cyclic amplification. On an average, each primer directs amplification of several discrete loci in the genome, making the assay useful for efficient screening of nucleotide sequence polymorphism between individuals. However, due to the stoichastic nature of DNA amplification with random sequence primers, it is important to optimise and maintain consistent reaction conditions for reproducible DNA amplification. RAPDs are DNA fragments amplified by the PCR using short synthetic primers (generally 10 bp) of random sequence.

These oligonucleotides serve as both forward and reverse primer and are usually able to amplify fragments from 1–10 genomic sites simultaneously. Amplified products (usually within the 0.5–5 kb size range) are separated on agarose gels in the presence of ethidium bromide and view under ultraviolet light and presence and absence of band will be observed.

These polymorphisms are considered to be primarily due to variation in the primer annealing sites, but they can also be generated by length differences in the amplified sequence between primer annealing sites. Each product is derived from a region of the genome that contains two short segments in inverted orientation, on opposite strands that are complementary to the primer. Kesseli and others compared the levels of polymorphism of two types of molecular markers, RFLP and RAPDs, as detected between two cultivars of lettuce in the construction of a genetic linkage map. RFLP and RAPD markers showed similar distributions throughout the genome, both identified similar levels of polymorphism. RAPD loci, however, were identified more rapidly.

21.5.1 Advantages and disadvantages

Advantages

The main advantage of RAPDs is that they are quick and easy to assay. Because PCR is involved, only low quantities of template DNA are required, usually 5–50 ng per reaction. Since random primers are commercially available, no sequence data for primer construction are needed. Moreover, RAPDs have a very high genomic abundance and are randomly distributed throughout the genome. They are dominant markers and hence have limitations in their use as markers for mapping, which can be overcome to some extent by selecting those markers that are linked in coupling.

RAPD assay has been used by several groups as efficient tools for identification of markers linked to agronomically important traits, which are introgressed during the development of near isogenic lines.

Disadvantages

The main drawback of RAPDs is their low reproducibility and hence highly standardised experimental procedures are needed because of their sensitivity to the reaction conditions. RAPD analyses generally require purified, high molecular weight DNA and precautions are needed to avoid contamination of DNA samples because short random primers are used that are able to amplify DNA fragments in a variety of organisms. Altogether, the inherent problems of reproducibility make RAPDs unsuitable markers for transference or comparison of results among research teams working in a similar species and subject. As for most other multilocus techniques, RAPD markers are not locus-specific, band profiles cannot be interpreted in terms of loci and alleles (dominance of markers) and similar sized fragments may not be homologous. RAPD markers were found to be easy to perform by different laboratories, but reproducibility was not achieved to a satisfactory level and, therefore, the method was utilised less for routine identifications. RAPD marker diversity was used also applied for diversity studies within and among some other Asteraceae species.

21.5.2 Applications

The application of RAPDs and their related modified markers in variability analysis and individual-specific genotyping has largely been carried out, but is less popular due to problems such as poor reproducibility faint or fuzzy products and difficulty in scoring bands, which lead to inappropriate inferences. RAPDs have been used for many purposes, ranging from studies at the individual level (e.g., genetic identity) to studies involving closely related species. RAPDs have also been applied in gene mapping studies to fill gaps not covered by other markers. Monteleone and others used this technique for the distinguish *mugo* and *uncinata* their subspecies. Variants of the RAPD technique include Arbitrarily Primed Polymerase Chain Reaction (AP-PCR), which uses longer arbitrary primers than RAPDs and DNA Amplification Fingerprinting (DAF) that uses shorter, 5–8 bp primers to generate a larger number of fragments. Multiple Arbitrary Amplicon Profiling (MAAP) is the collective term for techniques using single arbitrary primers.

21.6 Amplified fragment length polymorphism

Amplified fragment length polymerphism (AFLP), which is essentially intermediate between RFLPs and PCR. AFLP is based on a selectively amplifying a subset of restriction fragments from a complex mixture of DNA fragments obtained after digestion of genomic DNA with restriction endonucleases. Polymorphisms are detected from differences in the length of the amplified fragments by polyacrylamide gel electrophoresis (PAGE) or by capillary electrophoresis. The technique involves four steps: (i) restriction of DNA and ligation of oligonucletide adapters, (ii) preselective amplification, (iii) selective amplification and (iv) gel analysis of amplified fragments.

AFLP is a DNA fingerprinting technique, which detects DNA restriction fragments by means of PCR amplification. AFLP involves the restriction of genomic DNA, followed by ligation of adaptors complementary to the restriction sites and selective PCR amplification of a subset of the adapted restriction fragments. These fragments are viewed on denaturing polyacrylamide gels either through autoradiographic or fluorescence methodologies. AFLPs are DNA fragments (80–500 bp) obtained from digestion with restriction enzymes, followed by ligation of oligonucleotide adapters to the digestion products and selective amplification by the PCR. AFLPs therefore involve both RFLP and PCR. The PCR primers consist of a core sequence (part of the adapter) and a restriction enzyme specific sequence and 1–5 selective nucleotides (the higher the number of selective nucleotides, the lower the number of bands obtained per profile). The AFLP banding profiles are the result of variations in the restriction sites or in the intervening region.

The AFLP technique simultaneously generates fragments from many genomic sites (usually 50–100 fragments per reaction) that are separated by polyacrylamide gel electrophoresis and that are generally scored as dominant markers.

Selective Fragment Length Amplification (SFLA) and Selective Restriction Fragment Amplification (SRFA) are synonyms sometimes used to refer to AFLPs. A variation of the AFLP technique is known as Selectively Amplified Microsatellite Polymorphic Locus (SAMPL). Witsenboer and others studied the potential of SAMPL (Selectively Amplified Microsatellite Polymorphic Locus) analysis in lettuce to detect PCR-based codominant microsatellite markers. SAMPL is a method of amplifying microsatellite loci using general PCR primers. SAMPL analysis uses one AFLP primer in combination with a primer complementary to microsatellite sequences.

This technology amplifies microsatellite loci by using a single AFLP primer in combination with a primer complementary to compound microsatellite sequences, which do not require prior cloning and characterisation.

21.6.1 Advantages and disadvantages

Advantages

The strengths of AFLPs lie in their high genomic abundance, considerable reproducibility, the generation of many informative bands per reaction, their wide range of applications and the fact that no sequence data for primer construction are required. AFLPs may not be totally randomly distributed around the genome as clustering in certain genomic regions, such as centromers, has been reported for some crops. AFLPs can be analysed on automatic sequencers, but software problems concerning the scoring of AFLPs are encountered on some systems. The use of AFLP in genetic marker technologies has become the main tool due to its capability to disclose a high number of polymorphic markers by single reaction.

Disadvantages

Disadvantages include the need for purified, high molecular weight DNA, the dominance of alleles and the possible non-homology of comigrating fragments belonging to different loci. In addition, due to the high number and different intensity of bands per primer combination, there is the need to adopt certain strict but subjectively determined criteria for acceptance of bands in the analysis. Special attention should be paid to the fact that AFLP bands are not always independent. For example, in case of an insertion between two restriction sites the amplified DNA fragment results in increased band size. This will be interpreted as the loss of a small band and at the same time as the gain of a larger band.

This is important for the analysis of genetic relatedness, because it would enhance the weight of non-independent bands compared to the other bands. However, the major disadvantage of AFLP markers is that these are dominant markers.

21.6.2   Applications

AFLPs can be applied in studies involving genetic identity, parentage and identification of clones and cultivars and phylogenetic studies of closely related species because of the highly informative fingerprinting profiles generally obtained. Their high genomic abundance and generally random distribution throughout the genome make AFLPs a widely valued technology for gene mapping studies.

AFLP markers have successfully been used for analysing genetic diversity in some other plant species such as peanut, soyabean and maize. This technique is useful for breeders to accelerate plant improvement for a variety of criteria, by using molecular genetics maps to undertake marker-assisted selection and positional cloning for special characters.

Molecular markers are more reliable for genetic studies than morphological characteristics because the environment does not affect them. SAMPL is considered more applicable to intraspecific than to interspecific studies due to frequent null alleles. AFLP markers are useful in genetic studies, such as biodiversity evaluation, analysis of germplasm collections, genotyping of individuals and genetic distance analyses.

The availability of many different restriction enzymes and corresponding primer combinations provides a great deal of flexibility, enabling the direct manipulation of AFLP fragment generation for defined applications (e.g., polymorphism screening, QTL analysis, genetic mapping).

21.7 Minisatellites, variable number of tandem repeats (VNTR)

The term minisatellites was introduced by Jeffrey and others. These loci contain tandem repeats that vary in the number of repeat units between genotypes and are referred to as variable number of tandem repeats (VNTRs) (i.e., a single locus that contains variable number of tandem repeats between individuals) or hypervariable regions (HVRs) (i.e., numerous loci containing tandem repeats within a genome generating high levels of polymorphism between individuals). Minisatellites are a conceptually very different class of marker. They consist of chromosomal regions containing tandem repeat units of a 10–50 base motif, flanked by conserved DNA restriction sites. A minisatellite profile consisting of many bands, usually within a 4–20 kb size range, is generated by using common multilocus probes that are able to hybridise to minisatellite sequences in different species. Locus specific probes can be developed by molecular cloning of DNA restriction fragments, subsequent screening with a multilocus minisatellite probe and isolation of specific fragments. Variation in the number of repeat units, due to unequal crossing over or gene conversion, is considered to be the main cause of length polymorphisms. Due to the high mutation rate of minisatellites, the level of polymorphism is substantial, generally resulting in unique multilocus profiles for different individuals within a population.

21.7.1 Advantages and disadvantages

Advantages

The main advantages of minisatellites are their high level of polymorphism and high reproducibility.

Disadvantages

Disadvantages of minisatellites are similar to RFLPs due to the high similarity in methodological procedures. If multilocus probes are used, highly informative profiles are generally observed due to the generation of many informative bands per reaction. In that case, band profiles can not be interpreted in terms of loci and alleles and similar sized fragments may be non-homologous. In addition, the random distribution of minisatellites across the genome has been questioned.

21.7.2 Applications

The term DNA fingerprinting was introduced for minisatellites, though DNA fingerprinting is now used in a more general way to refer to a DNA-based assay to uniquely identify individuals. Minisatellites are particularly useful in studies involving genetic identity, parentage, clonal growth and structure and identification of varieties and cultivars and for population-level studies. Minisatellites are of reduced value for taxonomic studies because of hyper variability.

21.8 Polymerase chain reaction (PCR)-sequencing

The process of determining the order of the nucleotide bases along a DNA strand is called Sequencing. DNA sequencing enables us to perform a thorough analysis of DNA because it provides us with the most basic information of all, i.e., the exact order of the bases A, T, C and G in a segment of DNA. These methods are known as and the chemical degradation the chain termination method and were equally popular to begin with and even both teams shared the 1980 Nobel Prize, but Sanger's method became the standard because of its practicality.

PCR was a major breakthrough for molecular markers in that for the first time, any genomic region could be amplified and analysed in many individuals without the requirement for cloning and isolating large amounts of ultra-pure genomic DNA. PCR sequencing involves determination of the nucleotide sequence within a DNA fragment amplified by the PCR, using primers specific for a particular genomic site.

The method that has been most commonly used to determine nucleotide sequences is based on the termination of *in vitro* DNA replication.

21.8.1 Sanger's chain termination method

This method is based on the principle that single stranded DNA molecules that differ in length by just a single nucleotide can be separated from one another using polyacrylamide gel electrophoresis. The key to the method is the use of modified bases called Dideoxy nucleotide, due to which this method is also known as 'Sanger's Dideoxy sequencing method'. The dideoxy method gets its name from the critical role played by these synthetic nucleotides that lack the -OH at the 3′ carbon atom of De-oxy ribose sugar. A dideoxynucleotide-for ex-dideoxythymidine triphosphate or ddTTP can be added to the growing DNA strand but when, chain elongation stops as there is no 3′ -OH for the next nucleotide to be attached.

Hence, the dideoxy method is also called the chain termination method. The procedure is initiated by annealing a primer to the amplified DNA fragment, followed by dividing the mixture into four subsamples. Subsequently, DNA is replicated *in vitro* by adding the four deoxynucleotides (adenine, cytocine, guanine, thymidine; dA, dC, dG and dT), a single dideoxynucleotide (ddA, ddC, ddG or ddT) and the enzyme DNA polymerase to each reaction. Sequence extension occurs as long as deoxynucleotides are incorporated in the newly synthesised DNA strand. However, when a dideoxynucleotide is incorporated, DNA replication is terminated. Because each reaction contains many DNA molecules and incorporation of dideoxynucleotides occurs at random, each of the four subsamples contains fragments of varying length terminated at any occurrence of the particular dideoxy base used in the subsample. Finally, the fragments in each of the four subsamples are separated by gel electrophoresis.

21.8.2   Advantages and disadvantages

Advantages

Because all possible sequence differences within the amplified fragment can be resolved between individuals, PCR sequencing provides the ultimate measurement of genetic variation. Universal primer pairs to target specific sequences in a wide range of species are available for the chloroplast, mitochondria and ribosomal genomes.

Advantages of PCR sequencing include its high reproducibility and the fact that sequences of known identity are studied, increasing the chance of detecting truly homologous differences. Due to the amplification of fragments by PCR only low quantities of template DNA (the 'target' DNA used for the initial reaction) are required, e.g., 10–100 ng per reaction. Moreover, most of the technical procedures are amenable to automation.

Disadvantages

Disadvantages include low genome coverage and low levels of variation below the species level. In the event that primers for a genomic region of interest are unavailable, high development costs are involved. If sequences are visualised by polyacrylamide gel electrophoresis and autoradiography, analytical procedures are laborious and technically demanding. Fluorescent detection systems and reliable analytical software to score base pairs using automated sequencers are now widely applied.

This requires considerable investments for equipment or substantial costs in the case of outsourcing. Because sequencing is costly and time-consuming, most studies have focused on only one or a few loci. This restricts genome coverage and together with the fact that different genes may evolve at different rates, the extent to which the estimated gene diversity reflects overall genetic diversity is yet to be determined.

21.8.3   Applications

In general, insufficient nucleotide variation is detected below the species level and PCR sequencing is most useful to address questions of interspecific and intergeneric relationships. Until recently, chloroplast DNA and nuclear ribosomal DNA have provided the major datasets for phylogenetic inference because of the ease of obtaining data due to high copy number. Recently, single- to low-copy nuclear DNA markers have been developed as powerful new tools for phylogenetic analyses. Low-copy nuclear markers generally circumvent problems of uniparental inheritance frequently found in plastid markers and concerted evolution found in nuclear ribosomal DNA that limits their utility and reliability in phylogenetic studies. In addition to biparental inheritance, low-copy nuclear markers exhibit higher rates of evolution (particularly in intron regions) than cpDNA and nrDNA markers making them useful for closely related species. Yet another advantage is that lowcopy sequences generally evolve independently of paralogous sequences and tend to be stable in position and copy number.

21.9   Microsatellites or simple sequence repeat (SSR)

The term microsatellite is also known as Simple Sequence Repeats (SSRs), are sections of DNA, consisting of tandemly repeating mono-, di-, tri-, tetra- or penta-nucleotide units that are arranged throughout the genomes of most eukaryotic species. Microsatellite markers, developed from genomic libraries, can belong to either the transcribed region or the non transcribed region of the genome and rarely is there information available regarding their functions. Microsatellite sequences are especially suited to distinguish closely related genotypes; because of their high degree of variability, they are, therefore, favoured in population studies and for the identification of closely related cultivars.

Microsatellite polymorphism can be detected by Southern hybridisation or PCR. Microsatellites, like minisatellites, represent tandem repeats, but their repeat motifs are shorter (1–6 base pairs). If nucleotide sequences in the flanking regions of the microsatellite are known, specific primers (generally 20–25 bp) can be designed to amplify the microsatellite by PCR. Microsatellites and their flanking sequences can be identified by constructing a small-insert genomic library, screening the library with a synthetically labelled oligo-nucleotide repeat and sequencing the positive clones.

Alternatively, microsatellite may be identified by screening sequence databases for microsatellite sequence motifs from which adjacent primers may then be designed. In addition, primers may be used that have already been designed for closely related species. Polymerase slippage during DNA replication, or slipped strand mispairing, is considered to be the main cause of variation in the number of repeat units of a microsatellite, resulting in length polymorphisms that can be detected by gel electrophoresis. Other causes have also been reported.

21.9.1   Advantages and disadvantages

Advantages

The strengths of microsatellites include the codominance of alleles, their high genomic abundance in eukaryotes and their random distribution throughout the genome, with preferential association in low-copy regions. Because the technique is PCR-based, only low quantities of template DNA (10–100 ng per reaction) are required. Due to the use of long PCR primers, the reproducibility of microsatellites is high and analyses do not require high quality DNA. Although microsatellite analysis is, in principle, a single-locus technique, multiple microsatellites may be multiplexed during PCR or gel electrophoresis if the size ranges of the alleles of different loci do not overlap. This decreases significantly the analytical costs. Furthermore, the screening of microsatellite variation can be automated, if the use of automatic sequencers is an option EST-SSR markers are one class of marker that can contribute to 'direct allele selection', if they are shown to be completely associated or even responsible for a targeted trait. Yu and others identified two EST-SSR markers linked to the photoperiod response gene (ppd) in wheat. In recent years, the EST-SSR loci have been integrated, or genome-wide genetic maps have been prepared, in several plant (mainly cereal) species. A large number of genic SSRs have been placed on the genetic maps of wheat. Microsatellites can also be implemented as monolocus, codominant markers by converting individual microsatellite loci into PCR-based markers by designing primers from unique sequences flanking the microsatellite.

Microsatellite containing genomic fragment have to be cloned and sequenced in order to design primers for specific PCR amplification. This approach was called sequence-tagged microsatellite site (STMS). In the longer term, development of allele-specific markers for the genes controlling agronomic traits will be important for advancing the science of plant breeding. In this context, genic microsatellites are but one class of marker that can be deployed, along with single nucleotide polymorphisms and other types of markers that target functional polymorphisms within genes. The choice of the most appropriate marker system needs to be decided upon on a case by case basis and will depend on many issues, including the availability of technology platforms, costs for marker development, species transferability, information content and ease of documentation.

Disadvantages

One of the main drawbacks of microsatellites is that high development costs are involved if adequate primer sequences for the species of interest are unavailable, making them difficult to apply to unstudied groups. Although microsatellites are in principle codominant markers, mutations in the primer annealing sites may result in the occurrence of null alleles (no amplification of the intended PCR product), which may lead to errors in genotype scoring. The potential presence of null alleles increases with the use of microsatellite primers generated from germplasm unrelated to the species used to generate the microsatellite primers (poor 'crossspecies amplification'). Null alleles may result in a biased estimate of the allelic and genotypic frequencies and an underestimation of heterozygosity. Furthermore, the underlying mutation model of microsatellites (infinite allele model or stepwise mutation model) is still under debate. Homoplasy may occur at microsatellite loci due to different forward and backward mutations, which may cause underestimation of genetic divergence. A very common observation in microsatellite analysis is the appearance of stutter bands that are artifacts in the technique that occur by DNA slippage during PCR amplification. These can complicate the interpretation of the band profiles because size determination of the fragments is more difficult and heterozygotes may be confused with homozygotes. However, the interpretation may be clarified by including appropriate reference genotypes of known band sizes in the experiment.

21.9.2   Applications

In general, microsatellites show a high level of polymorphism. As a consequence, they are very informative markers that can be used for many population genetics studies, ranging from the individual level (e.g., clone and strain identification) to that of closely related species. Conversely, their high mutation rate makes them unsuitable for studies involving higher taxonomic levels. Microsatellites are also considered ideal markers in gene mapping studies. Molecular markers have proven useful for assessment of genetic variation in germplasm collections. Expansion and contraction of SSR repeats in genes of known function can be tested for association with phenotypic variation or, more desirably, biological function. Several studies have found that genic SSRs are useful for estimating genetic relationship and at the same time provide opportunities to examine functional diversity in relation to adaptive variation.

21.10   Inter simple sequence repeats (ISSR)

ISSRs are DNA fragments of about 100–3000 bp located between adjacent, oppositely oriented microsatellite regions. This technique, reported by Zietkiewicz and others primers based on microsatellites are utilised to amplify inter-SSR DNA sequences. ISSRs are amplified by PCR using microsatellite core sequences as primers with a few selective nucleotides as anchors into the non-repeat adjacent regions (16–18 bp). About 10–60 fragments from multiple loci are generated simultaneously, separated by gel electrophoresis and scored as the presence or absence of fragments of particular size. Techniques related to ISSR analysis are Single Primer Amplification Reaction (SPAR) that uses a single primer containing only the core motif of a microsatellite and Directed Amplification of Minisatellite region DNA (DAMD) that uses a single primer containing only the core motif of a minisatellite.

21.10.1 Advantages and disadvantages

Advantages

The main advantage of ISSRs is that no sequence data for primer construction are needed. Because the analytical procedures include PCR, only low quantities of template DNA are required (5–50 ng per reaction). Furthermore, ISSRs are randomly distributed throughout the genome. This is mostly dominant marker, though occasionally its exhibits as codominance.

Disadvantages

Because ISSR is a multilocus technique; disadvantages include the possible nonhomology of similar sized fragments. Moreover, ISSRs, like RAPDs, can have reproducibility problems.

21.10.2 Applications

Because of the multilocus fingerprinting profiles obtained, ISSR analysis can be applied in studies involving genetic identity, parentage, clone and strain identification and taxonomic studies of closely related species. In addition, ISSRs are considered useful in gene mapping studies.

21.11 Single-strand conformation polymorphism (SSCP)

SSCPs are DNA fragments of about 200–800 bp amplified by PCR using specific primers of 20–25 bp. Gel electrophoresis of single-strand DNA is used to detect nucleotide sequence variation among the amplified fragments. The method is based on the fact that the electrophoretic mobility of singlestrand DNA depends on the secondary structure (conformation) of the molecule, which is changed significantly with mutation. Thus, SSCP provides a method to detect nucleotide variation among DNA samples without having to perform sequence reactions. In SSCP the amplified DNA is first denatured and then subject to non-denaturing gel electrophoresis. Related techniques to SSCP are Denaturing Gradient Gel Electrophoresis (DGGE) that uses double stranded DNA which is converted to single stranded DNA in an increasingly denaturing physical environment during gel electrophoresis and Thermal Gradient Gel

Electrophoresis (TGGE) which uses temperature gradients to denature double stranded DNA during electrophoresis.

21.11.1 Advantages and disadvantages

Advantages

Advantages of SSCP are the codominance of alleles and the low quantities of template DNA required (10–100 ng per reaction) due to the fact that the technique is PCR-based.

Disadvantages

Drawbacks include the need for sequence data to design PCR primers and the necessity of highly standardised electrophoretic conditions in order to obtain reproducible results. Furthermore, some mutations may remain undetected and hence absence of mutation cannot be proven.

21.11.2 Applications

SSCPs have been used to detect mutations in genes using gene sequence information for primer construction.

21.12 Cleaved amplified polymorphic sequence (CAPS)

CAPS are DNA fragments amplified by PCR using specific 20–25 bp primers, followed by digestion of the PCR products with a restriction enzyme. Subsequently, length polymorphisms resulting from variation in the occurrence of restriction sites are identified by gel electrophoresis of the digested products. CAPS have also been referred to as PCR-Restriction Fragment Length Polymorphism (PCR-RFLP).

21.12.1 Advantages and disadvantages

Advantages

Advantages of CAPS include the involvement of PCR requiring only low quantities of template DNA (50–100 ng per reaction), the codominance of alleles and the high reproducibility. Compared to RFLPs, CAPS analysis does not include the laborious and technically demanding steps of Southern blot hybridisation and radioactive detection procedures. These markers are codominant in nature.

Disadvantages

In comparison with RFLP analysis, CAPS polymorphisms are more difficult to find because of the limited size of the amplified fragments (300–1800 bp). Sequence data needed for synthesis of the primers.

21.12.2 Applications

CAPS markers have been applied predominantly in gene mapping studies.

21.13   Sequence characterised amplified region (SCAR)

Martin and others introduced this technique wherein the RAPD marker termini are sequenced and longer primers are designed (22–24 nucleotide bases long) for specific amplification of a particular locus. SCARs are DNA fragments amplified by the PCR using specific 15–30 bp primers, designed from nucleotide sequences established from cloned RAPD fragments linked to a trait of interest. By using longer PCR primers, SCARs do not face the problem of low reproducibility generally encountered with RAPDs. Obtaining a codominant marker may be an additional advantage of converting RAPDs into SCARs, although SCARs may exhibit dominance when one or both primers partially overlap the site of sequence variation. Length polymorphisms are detected by gel electrophoresis.

21.13.1 Advantages and disadvantages

Advantages

The main advantage of SCARs is that they are quick and easy to use. In addition, SCARs have a high reproducibility and are locus-specific. Due to the use of PCR, only low quantities of template DNA are required (10–100 ng per reaction).

Disadvantages

Disadvantages include the need for sequence data to design the PCR primers.

21.13.2 Applications

SCARs are locus specific and have been applied in gene mapping studies and marker assisted selection.

21.14   Single nucleotide polymorphism (SNP)

A noval class of DNA markers namely single nucleotide polymorphism in genome (SNPs) has recently become highly proffered in genomic studies. The fact that in many organisms most polymorphisms result from changes in a single nucleotide position (point mutations), has led to the development of techniques to study single nucleotide polymorphisms (SNPs). Analytical procedures require sequence information for the design of allelespecific PCR primers or oligonucleotide probes. SNPs and flanking sequences can be found by library construction and sequencing or through the screening of readily available sequence databases. Once the location of SNPs is identified and appropriate primers designed, one of the advantages they offer is the possibility of high throughput automation. To achieve high sample throughput, multiplex PCR and hybridisation to oligonucleotide microarrays or analysis on automated sequencers are often used to interrogate the presence of SNPs. SNP analysis may be useful for cultivar discrimination in crops where it is difficult to find polymorphisms, such as in the cultivated tomato. SNPs may also be used to saturate linkage maps in order to locate relevant traits in the genome. For instance, in *Arabidopsis thaliana* a highdensity linkage map for easy to score DNAmarkers was lacking until SNPs became available.

To date, SNP markers are not yet routinely applied in genebanks, in particular because of the high costs involved. Retrotransposon-based markers Retrotransposons consist of long terminal repeats (LTR) with a highly conserved terminus, which is exploited for primer design in the development of retrotransposon-based markers. Retrotransposons have been found to comprise the most common class of transposable elements in eukaryotes and to occur in high copy number in plant genomes. Several of these elements have been sequenced and were found to display a high degree of heterogeneity and insertional polymorphism, both within and between species. Because retrotransposon insertions are irreversible, they are considered particularly useful in phylogenetic studies.

In addition, their widespread occurrence throughout the genome can be exploited in gene mapping studies and they are frequently observed in regions adjacent to known plant genes. Several variations of retrotransposon-based markers exist. Sequence-Specific Amplified Polymorphism (S-SAP) is a dominant, multiplex marker system for the detection of variation in DNA flanking the retrotransposon insertion site. Retrotransposon containing fragments are amplified by PCR, using one primer designed from the conserved terminus of the LTR and one based on the presence of a nearby restriction endonucleases site. Experimental procedures resemble those used for AFLP analysis and they are usually dominant markers.

Compared to AFLP, SSAP generally yields fewer fragments but higher levels of polymorphism. Interretrotransposon Amplified Polymorphism (IRAP) and Retrotransposon- Microsatellite Amplified Polymorphism (REMAP) are dominant, multiplex marker systems that examine variation in retrotransposon insertion sites. With IRAP, fragments between two retrotransposons are isolated by PCR, using outward facing primers annealing to LTR target sequences. In the case of REMAP, fragments between retrotransposons and microsatellites are amplified by PCR, using one primer based on a LTR target sequence and one based on a simple sequence repeat motif. IRAP as well as REMAP fragments can be separated by high-resolution agarose gel electrophoresis. Retrotransposon- Based Insertional Polymorphism (RBIP) is a codominant marker system that uses PCR primers designed from the retrotransposon and its flanking DNA to examine insertional polymorphisms for individual retrotransposons. Presence or absence of insertion is investigated by two PCRs, the first using one primer from the retrotransposon and one from the flanking DNA, the second using primers designed from both flanking regions. Polymorphisms are detected by simple agarose gel electrophoresis or by dot hybridisation assays. A drawback of the method is that sequence data of the flanking regions is required for primer design.

Comparative qualities of marker techniques: DNA provides many advantages that make it especially attractive in studies of diversity and relationships. These advantages have included: (i) freedom from environmental and pleiotropic effects. Molecular markers do not exhibit phenotypic plasticity, while morphological and biochemical markers can vary in different environments. DNA characters have a much better chance of providing homologous traits. Most morphological or biochemical markers, in contrast, are under polygenic control and subject to epistatic control and environmental modification (plasticity), (ii) a potentially unlimited number of independent markers are available, unlike morphological or biochemical data, (iii) DNA characters can be more easily scored as discrete states of alleles or DNA base pairs, while some morphological, biochemical and field evaluation data must be scored as continuously variable characters that are less amenable to robust analytical methods, (iv) many molecular markers are selectively neutral. These advantages do not imply that other more traditional data used to characterise biodiversity are not valuable.

On the contrary, morphological, ecological and other 'traditional' data will continue to provide practical and often critical information needed to characterise genetic resources. Molecular markers differ in many qualities and must therefore be carefully chosen and analysed differently with their differences in mind. To assist in choosing the appropriate marker technique, an overview of the main properties of the marker technologies described in Table 21.1.

Table 21.1: Summary advantage and disadvantage of some commonly used markers.

Type of markers	Advantages	Disadvantages
Restriction fragment Length polymorphism (RFLP)	High genomic abundance Co-dominant markers Highly reproducible Can use filters many times Good genome coverage Can be used across species No sequence information	Need large amount of good quality DNA Laborious (compared to RAPD) Difficult to automate Need radioactive labelling Cloning and characterisation of probe are required

(Cont'd...)

Type of markers	Advantages	Disadvantages
	Can be used in plants reliably (well-tested) Needed for map based cloning	
Randomly amplified Polymorphic DNA (RAPD)	High genomic abundance Good genome coverage No sequence information Ideal for automation Less amount of DNA (poor DNA acceptable) No radioactive labelling Relatively faster	No probe or primer information Dominant markers Not reproducible Can not be used across species Not very well-tested
Simple sequence repeat (SSR)	High genomic abundance Highly reproducible Fairly good genome coverage High polymorphism No radioactive labelling Easy to automate Multiple alleles	Can not be used across species Need sequence information Not well-tested
Amplified fragment length polymorphism (AFLP)	High genomic abundance High polymorphism No need for sequence information Can be used across species Work with smaller RFLP fragments Useful in preparing contig maps	Very tricky due to changes in patterns with respect to materials used Cannot get consistent map (not reproducible) Need to have very good primers
Sequence-tagged site (STS)	Useful in preparing contig maps No radioactive labelling Fairly good genome coverage Highly reproducible Can use filters many times	Laborious Cannot detect mutations out of the target sites Need sequence information Cloning and characterisation of probe are required
Isozymes	Useful for evolutionary studies Isolation lot easier than that of DNA Can be used across species No radioactive labelling No need for sequence information	Laborious Limited in polymorphism Expensive (each system is unique) Have to know the location of the tissue -not easily automated

21.14.1 Genomic abundance

The number of markers that can be generated is determined mainly by the frequency at which the sites of interest occur within the genome. RFLPs and

AFLPs generate abundant markers due to the large number of restriction enzymes available and the frequent occurrence of their recognition sites within genomes. Within eukaryotic genomes, microsatellites have also been found to occur frequently. RAPD markers are even more abundant because numerous random sequences can be used for primer construction. In contrast, the number of allozyme markers is restricted due to the limited number (about 30) of enzyme detection systems available for analysis. To investigate specific genomic regions by PCR sequencing, SSCP, CAPS or SCAR, sequence data of the sites of interest (structural genes mainly) are required for primer construction. Although, in principle, many sites of interest may occur within genomes, the proportion of the genome covered by PCR sequencing, SSCP, CAPS and SCAR in studies reported to date is limited. However, this is expected to change due to the wealth of sequence information that is becoming increasingly available for different crops. Genomic abundance is essential to studies where a large fraction of the genome needs to be covered, e.g., for the development of high-density linkage maps in gene mapping studies. If, in addition to genomic abundance, genome coverage is also sought, caution should be taken in marker selection. While some markers are known to be scattered quite evenly across the genomes, others, such as some AFLP markers, sometimes cluster in certain genomic regions. For example, clustering of AFLP markers has been reported in centromeric regions of *Arabidopsis thaliana*, soyabean and rye.

Level of polymorphism: The resolving power of genetic markers is determined by the level of polymorphism detected, which is determined by the mutation rate at the genomic sites involved. Variation at allozyme loci is caused by point mutations, which occur at low frequency (<10–6 per meiosis). Moreover, only mutations modifying the net electric charge and conformation of proteins can be detected, reducing the resolving power of allozymes.

The other markers generally show intermediate levels of polymorphism, resulting from base substitutions, insertions or deletions which may alter primer annealing sites and recognition sites of restriction enzymes, or change the size of restriction fragments and amplified products. In choosing the appropriate technique, the level of polymorphism detected by the marker needs to be considered in relation to the presumed degree of genetic relatedness within the material to be studied. Higher resolving power is required when samples are more closely related. For example, analyses within species or among closely related species may call for fast evolving markers such as microsatellites. However if the objective is to study genetic relatedness at higher taxonomic levels (such as congeneric species), AFLPs or RFLPs may be a better choice because comigrating fast-evolving markers will have less chance of being homologous. A primary guiding principle in marker selection is that more conservative markers (those having slower evolutionary rates) are needed with increasing evolutionary distance and vice-versa.

Locus-specificity: Genetic markers using multi locus probes or primers benefit from the fact that multiple polymorphisms, representing various genomic regions, are generated simultaneously. However, a major drawback is that in general the band profiles cannot be interpreted in terms of loci and alleles, but are scored as the presence or absence of bands of a particular size. As a consequence, similar sized fragments may represent alleles from different loci and not be homologous. Therefore, locus-specific markers should be considered for questions of phylogeny or genetic relatedness. Alternatively, markers for fingerprinting studies rely on differences only and homology is not a concern. In general, locus-specific markers generate polymorphisms of known identity, however in most cases sequencing data are needed for their development.

Codominance of alleles: Codominant markers are markers for which both alleles are expressed when co-occurring in an individual. Therefore, with codominant markers, heterozygotes can be distinguished from homozygotes, allowing the determination of genotypes and allele frequencies at loci. In contrast, band profiles of dominant markers are scored as the presence or absence of fragments of a particular size and heterozygosity cannot be determined directly.

As a consequence, only an approximation of allele frequency can be obtained by assuming Hardy-Weinberg equilibrium in a population and estimating allele frequency from the proportion of individuals with the absent phenotype (homozygous recessive). For predominantly self-fertilising species, heterozygosity could be disregarded and allele frequencies be considered equal to observed band frequencies. Codominant markers are preferred for most applications. The majority of codominant markers are single locus markers and hence the degree of information per assay is usually lower compared to the multilocus techniques.

Reproducibility: Reproducibility is always an important property of markers, but even more important with collaborative projects, involving the generation of data by different labs whose results need to be assembled. To obtain reproducible results, the extraction of purified, high quality DNA is a prerequisite for the majority of the marker techniques. For example, degraded and/or unpurified DNA may affect the amplification or restriction of DNA, resulting in unspecific polymorphisms. Even when purified and high molecular weight DNA is used, RAPDs often fail to show reproducible results. This is because RAPD primers are very short (10 bp), which can result in alterations in their annealing behaviours to the template DNA and the resulting band profiles as a result of small deviations in experimental conditions. Therefore, highly standardised experimental procedures are required when RAPD markers are being used. This implies the need for including repeated samples and also the inclusion of reference genotypes that represent bands of known size. Problems with reproducibility in RAPD analysis could be overcome by focusing on mapped markers for which their inheritance has already been verified.

Labour-intensity: RFLPs and minisatellites are labour-intensive markers because their analysis includes the time-consuming steps of Southern blotting, labelling of probes and hybridisation. Therefore, PCR based techniques are currently preferred, some of which can even be automated to decrease the labour-intensity. PCR sequencing may still be quite labour-intensive if performed by the old time consuming method of performing four separate sequence reactions per sample. However, automated procedures have greatly reduced labour-intensity of PCR-sequencing. The labour-intensity of the other PCR-based techniques presented varies from low to medium, depending on the methodological procedures required in addition to PCR.

Technical demands: RFLPs, minisatellites and manual PCR sequencing require higher technical skills and facilities for analysis. RFLP and minisatellite analyses require Southern blot hybridisations and may include radioactive labelling. This calls for expertise and exclusive facilities needed to comply with special legal and safety requirements. These technologies are therefore among the most technically demanding markers. Another type of technical demand arises from the use of polyacrylamide gels and automated equipment. Allozymes and PCR-based markers analysed on agarose gels (e.g., RAPD, SCAR and microsatellites) are the least technically demanding.

Operational costs: Wages, laboratory facilities, technical equipment and consumables all contribute to the operational costs of the technologies. Relatively expensive consumables include Taqpolymerase needed for all PCR based marker types, restriction enzymes (for RFLPs, minisatellites and CAPS and particularly the restriction enzyme MseI often used in AFLPs) and isotopes where polymorphisms are visualised by means of radioactive labelling. Polyacrylamide gels are more expensive to run than agarose gels and require visualisation of polymorphisms by autoradiography or silver staining procedures, which are more costly compared to ethidium-bromide staining. Laborious and technically demanding markers, such as RFLPs, minisatellites, PCR sequencing and those techniques being performed by automated equipment, are quite expensive. Costs of performing RAPD analyses are usually considered low. However, if measures to ensure reproducibility and low numbers of markers per primer are taken into account, costs may increase to the level of the more complex technologies. In general, operational costs of markers will vary depending on the methodology. Regarding automated procedures and technologies, while purchasing the equipment is usually very expensive and the technical expertise required is high, a significant increase in throughput may be obtained through multiplexing. An additional consideration is the emergence of cost effective 'outsourcing' companies to generate marker-based and DNA sequencing data, as service laboratories keep up with efficient equipment developments. Outsourcing allows researchers to concentrate on defining questions, experimental design, data analysis and interpretation. The relative costs/benefits of outsourcing will vary in different labs according to local labour and supply costs, availability of equipment, the benefit of generating your own data for quality control or educational purposes and the legal requirements to ship crop germplasm DNA out of a country.

Development costs: Marker development may be very time-consuming and costly when suitable probes or sequence data for primer construction are unavailable. Development of suitable probes for Southern blot hybridisations (e.g., for RFLP analysis) requires the construction of either genomic or cDNA libraries and the examination of various probe/restriction enzyme combinations for their ability to detect polymorphisms. The development of site-specific PCR primers (e.g., for microsatellite analysis) also requires the construction of libraries, which then need to be screened to identify the fragments of interest. Subsequently, the identified fragments need to be sequenced to verify their suitability and to design primers. Therefore, the investment required for marker development should be evaluated in relation to the intended range of application of the technique. Alternatively, new genomic tools are allowing probes, primers and sequence data to be obtained from genome databases of other species, with the understanding, as in all DNA tools, that their usefulness may decrease with increasing evolutionary distance between the species.

Quantity of DNA required: Because only small quantities of template DNA (5–100 ng per reaction) are required, techniques, which are based on the PCR, are currently preferred. Although RFLPs and minisatellites require the largest amount of DNA (5–10 μg per reaction), Southern blot membranes may be probed several times. Intermediate quantities of DNA are needed for AFLP-analysis (0.3–1 μg per reaction) because restriction of the DNA precedes the PCR reaction. In general, consideration should be given to the use of PCR-based markers if only small amounts of DNA can be obtained.

Amenability to automation: Currently, if adequate equipment and resources are available, techniques that can be automated are highly preferred because of the potential for high sample throughput. Although considerable financial investment is still required, automation may be cost effective when techniques are applied on a routine basis. As pointed out above, outsourcing of data generation may also be an alternative strategy. Nearly all techniques that are based on the PCR are amenable to a certain degree of automation.

Mobile DNA

22.1 Introduction

Mobile DNA consists of blocks of DNA that are able to move and insert into new locations throughout the genome without needing DNA sequence similarity or requiring the process of homologous recombination to enable movement. Such a process of movement and insertion is often called transposition, and the mobile DNAs are frequently called transposons. Some mobile DNAs (DNA elements) such as retrotransposons move by different mechanisms than used by strictly defined transposons, but achieve the same overall outcome - namely movement of blocks of DNA to new positions in chromosomes. The general topic of mobile DNA also encompasses other genetic entities that are able to move between organisms such as plasmids and viruses, which in some situations reside within cells permanently as an extra, optional, replicating chromosome. Plasmids and bacterial viruses (bacteriophages) are involved in much movement of mobile DNA between different micro-organisms, and plasmids often carry a variety of different transposons.

22.2 Transposition

Transposition is a process where sections of DNA move throughout the genome, often making copies of themselves at the same time. Not all DNA does this, only sections called transposons, transposable elements or jumping genes. Transposons have been called junk DNA, because they have no known useful function, and selfish DNA because they exploit the genetic mechanisms of the cell. A more modern view is that these terms do not accurately convey the full range of transposon behaviour, which ranges from parasitism to mutualism. Their activities have conferred upon pathogenic bacteria the disturbing advantage of acquiring resistance to a number of antibiotics, and then transmitting this multiple drug resistance in a highly infectious way. Increasingly transposons are being seen to participate in evolution of new function and complex regulatory circuits in multicellular organisms.

22.2.1 Mechanism of transposition

Transposons usually code for at least one gene, as they need the enzyme transposase, and the transposon usually codes for this itself. The ends of the transposon consist of terminal DNA repeats which the transposase binds to.

The transposase also binds to the target site on the genome, which is cut to leave 'sticky ends', or single stranded DNA overhangs similar to those created by restriction endonucleasess. Depending on the type of transposon it is then either copied or cut out and inserted into the target site by the action of the enzyme DNA ligase. Some transposons encode mechanisms for copying themselves, but many exploit the cells own DNA copying mechanisms.

22.2.2 Types of transposon

There are three different types of transposon, classified by their mechanisms:

Class I: retrotransposons

Retrotransposons (sometimes Retroposons) work by copying themselves and pasting copies back into the genome, in multiple places. First retrotransposons copy themselves to RNA (transcription), but instead of being translated the RNA is picked up by reverse transcriptase (often coded by the transposon itself) which copies the RNA back into DNA which is then inserted into the genome. This process is called retroposition, and the new genes are called retrogenes. Retrotransposons usually have very long terminal repeats of over 1000 base pairs, making them easy to find and study. Retrotransposons behave very similarly to retroviruses, such as HIV, giving a clue to their evolutionary origins.

Retrotransposons can carry flanking sequence during their mobilisation to a new site, a process which is called 3'-transduction, and which can cause duplication of genes. Such gene duplication is an important mechanisms for creating new genes and generating genomic novelty. The duplication of entire genes and the creation of previously undescribed gene families through retrotransposon-mediated flanking DNA sequence transduction has been highlighted recently as an important mechanism by which mobile DNA impacts the host genome during evolution.

Class II transposons

Class II transposons move by cut and paste, rather than copy and paste, using the transposase enzyme. Different types of transposase work in different ways. Some can bind to any part of the DNA molecule, and the target site can therefore be anywhere, while others bind to specific sequences. The transposase then cuts the target site to produce sticky ends, cuts out the transposon and ligases it into the target site, and then fills in the sticky ends with their base pairs.

Transposons may lose their ability to synthesise transposase through mutation, yet continue to jump through the genome because other transposons are still producing transposase. It is possible that some transposons become transposons simply by mutations producing the sequences that signals to transposase.

P elements: P elements are a particular type of class II transposon that was first detected in Drosophila (fruit flies) in the 1950s. P elements are so good at spreading that they cause so many mutations in the germ line cells as to sterilise the flies. *Drosophila* have evolved a mechanism for suppressing transposase, however, and all except a few isolated lab populations are immune to the effects of P elements. The roundworm *Caenorhabditis elegans* has evolved an RNA interference system to respond to a similar problem of transposons.

22.2.3   MULEs and PACK-MULEs

The Mutator (Mu) transposons of maize are known to capture pieces of host genes and then rapidly amplify them via subsequent duplications. Related Mu Like Elements (MULEs) are a widely distributed plant and fungal transposon family that are thought to be the result of this process.

Similar examples of gene capture have been observed in other class II elements such as CACTA elements, hAT elements and Helitrons, suggesting that this gene capture process may be a feature of many transposons.

MULEs have terminal inverted repeats (TIRs) and the inverted repeats (TIRs) from MULEs often flank the captured fragments of host genes. Such TIRs of MULEs with captured genes are referred to as Pack-MULEs.

There are over 3,000 Pack-MULEs in rice containing fragments derived from more than 1,000 cellular genes. Pack-MULEs frequently contain fragments from multiple genes that are fused to form new genes. Some captured gene fragments in rice PACK-MULES may be functional new genes.

Class III: Miniature inverted-repeat transposable elements

MITEs are sequences of about 400 base pairs and 15 base pair inverted repeats that vary very little. They are found in their thousands in the genomes of both plants and animals (over 100,000 were found in the rice genome). MITEs are too small to encode any proteins.

22.2.4   Effects of transposons on the genome

Transposons can cause mutation. They do this either by being inserted into the coding region of a gene, losing some of the gene in the process, or by being inserted upstream of the coding region of a gene in an area important in determining the expression of the gene, such an area where a transcription factor would bind to the DNA. Insertion into a gene can also cause alternative splicing leading to a greater diversity of proteins produced by a single gene.

Both transposons and retrotransposons can cause segments of DNA to be duplicated, an important first stage in the evolution of new information, and a process that lead more rapid sequence change in the duplicated genes which are less constrained by purifying selection.

Retrotransposons have been shown to be active in promoting much gene evolution in primate genomes, including the human genome - by creating new genes and generating genomic novelty. For instance, SVA is the youngest retrotransposon family in primates is able to capture 3′ flanking sequences during retrotransposition (that is carry out 3′-transduction to effectively duplicate genes).

Recent study of the human genome the human genome has shown that about 53 thousand base-pairs of genomic sequences have been duplicated by 143 different SVA-mediated flanking gene capture (3′ transduction) events.

One group of SVA elements that duplicated the entire AMAC gene three times in the human genome which happened before the divergence of humans and African great apes. In addition to the original AMAC gene, the three transduced (duplicated) AMAC copies contain intact protein coding sequences in the human genome, and at least two are actively transcribed in different human tissues.

Transposons are involved the creation of new 'chimeric' (hybrid) proteins by gene fusion , and in generation of the circuitry of complex new gene regulatory networks during natural evolution. An important illustration of this evolutionary role has been found in human evolution, involving a member of the mariner-family transposons called Hsmar1. Hsmar1 (Homo sapiens mariner) is present in the human genome as around 1500 copies of mobile DNA. SETMAR, a new primate chimeric gene resulting from fusion of a SET histone methyltransferase gene to the transposase gene of Hsmar1 mobile DNA. The transposase gene was recruited as part of SETMAR 40–58 million years ago in an anthropoid ape, after the insertion of an Hsmar1 transposon downstream of a preexisting SET gene.

Transposition of genes to new chromosomal locations via retrotransposons (retroposition) has generated a significant number of new functional genes (retrogenes) in mammalian and invertebrate animal genomes. A burst of retroposition in primates led to emergence of many new human genes during primate evolution over the last ~63 million years of primate evolution.

Transposons also increase the size of the genome, because they leave multiple copies of themselves in the genome. For instance, the maize genome has doubled in size in the past 3–6 million years, largely due to a massive burst of retrogene formation.

Transposons can play a part in causing diseases including cancer and haemophilias A (when the Factor VIII gene is disrupted) and B (Factor IX).

22.2.5   Evolution of transposons

The evolution of transposons is an area of much current research, and we can't yet say exactly how transposons came about. However, we do know that many similar transposons can be found in all major groups of organisms on earth. There are three plausible hypotheses as to how this came about:

- Transposons were present in the last universal common ancestor.
- Transposons have arisen a number of times independently (less likely when the transposons are so similar).
- Transposons can be spread in viruses or by other mechanisms of horizontal gene transfer.

The hypothesis that transposons can be spread in viruses is supported by evidence in bacteria. In many bacteria transposons contain genes for antibiotic resistance, and when snipped out of the bacteria's circular DNA the plasmids produced can be transferred to other bacterial cells, spreading the resistance to other species by horizontal gene transfer.

In terms of gene selection, there is an obvious selective advantage for bits of DNA that do nothing but copy themselves and allow the rest of the genome to do all the hard work, from the point of view of that piece DNA. At first sight it would like transposons are bad for the genone as a whole, jumping around causing mutations and silencing of genes, and that organisms would evolve defenses against these DNA parasites. It's possible, however, that transposons are actually selectively advantageous for the genome as a whole. The bulking out of the genome with transposons and other non-coding DNA could make gene regulation easier.

Some retrotransposons and retroviruses have sequences similar to the exon, promoter and enhancer regions and DNA, and this may also play a part in gene regulation, or provide a good evolutionary substrate. Transposon activity is greater when the organism is under stress, and this could lead to increased mutation rates when an organism is in an environment it is not well adapted to. There is also evidence that genomes are more stable and better able to survive temperature stress the larger they are.

Discovery of transposons

Transposons were first discovered in maize in 1948 by Barbara McClintock. McClintock noticed unusual genetic behaviour such as maize kernel variegation and genetic instability that she attributed to DNA insertion, deletion translocation, and somatic mutation events. Such observation were not new to science but McClintock's interpretation of it was 'before it's time'. Her discovery of mobile DNA was initially greeted with widespread skepticism as the interpretation of the strange behaviour did not fit with conventional genetics ideas of that time of that time. However, as more evidence for mobile DNA in micro-organisms, plant and insects became known, the significance of her discovery was widely recognised.

Some uses for transposons

Transposons can be exploited by scientists for studying genetics. Because transposons insert themselves into genes they can be used to knock out genes. This technique turns genes off so that their function can be determined, and has been experimentally exploited in micro-organisms, plants and animals.

22.3 Mobile DNA in obligate intracellular bacteria

The genome sciences are producing a wealth of new information on the abundance and distribution of mobile DNA in prokaryotes, examples of which include plasmids, bacteriophages and transposable elements. Findings so far indicate that most bacterial genomes harbour prophages, some of which occupy up to 20% of the host genome. In fact, mobile-related DNA, such as prophages, account for more than 50% of the strain-specific DNA in several important pathogens and are the most common transporters of virulence genes in bacteria. Plasmids and transposable elements have also had a considerable effect on bacterial genome architecture. Therefore, the importance of mobile DNA to our understanding of microbial genomes is profound.

This section discusses the data on mobile DNA in obligate intracellular bacteria, a group of organisms that has been characterised in the microbiology literature as having few if any mobile genes, owing to their intracellular confinement and accelerated rates of gene loss. It is this notion that makes the growing discovery of mobile genetic elements in these species surprising.

22.3.1 Obligate intracellular bacteria

The prokaryotes can be divided into three broad categories on the basis of their lifestyle: free-living, facultative intracellular and obligate intracellular bacteria. Free-living bacteria tend to have large population sizes and genomes (4–10 Mb) with a moderate composition of mobile DNA. Facultative intra cellular species, which are not confined to intracellular replication, tend to be pathogenic with intermediate population sizes and a genome size that is similar to free-living species (2–7 Mb). Obligate intracellular bacteria, also known as obligate endosymbionts, replicate exclusively inside the cells of mostly eukaryotic organisms and typically have no extracellular state. They tend to have small population sizes, and their genomes are usually small (0.5–2 Mb) and show marked AT nucleotide biases, accelerated sequence evolution and a loss of genes that are involved in recombination and repair pathways.

Obligate intracellular species can be further divided into species that show strict vertical transmission and species that show at least some horizontal transmission. The former includes the dietary endosymbionts that are required for the survival and reproduction of their insect hosts (for example, *Buchnera* spp. of aphids, *Wigglesworthia* spp. of tsetse flies, and *Blochmannia* spp. of ants). These genera of γ-proteobacteria manifest strict maternal inheritance, obligate mutualistic associations and precise COSPECIATION PATTERNS with their hosts. By contrast, species that show at least some horizontal transmission include human and plant pathogens (for example, *Chlamydia* spp., *Rickettsia* spp. and *Phytoplasma* spp.) and the reproductive parasites of arthropods (for example, *Wolbachia* spp. and *Candidatus Cardinium hertigii*) that can distort sex ratios and sex determination. Species that can switch from one host to another tend to form PARASITIC ASSOCIATIONS, but can also form MUTUALISTIC ASSOCIATIONS or COMMENSAL ASSOCIATIONS.

Comparative genomic studies of obligate intracellular species mostly reveal remarkable conservation in gene content, genome size and gene order. The impact of REDUCTIVE EVOLUTION facilitated by small population sizes, DELETION BIASES and constrained access to novel gene pools in these species might promote genome streamlining and a lack of horizontal gene transfer. Indeed, the published genomes of five endosymbiotic γ-proteobacteria of insects that are obligate mutualists, including the genomes of three *Buchnera* strains, one *Wigglesworthia* strain and one *Blochmannia* strain, are devoid of mobile genetic elements. Furthermore, comparisons between the genomes of different species of *Buchnera* indicate that there has been no gene influx (duplications or horizontal-gene-transfer events) over the past 50 million years. For these reasons, obligate intracellular species are commonly presumed to be impervious to mobile genetic elements and, for the long-term, vertically transmitted obligate mutualists, this presumption is probably correct. However, the study of species that can switch hosts will probably provide the greatest insights and will facilitate tractable hypotheses on the evolution of mobile DNA and genome instability in obligate intracellular bacteria. Here, we summarise the evidence that amends the view that obligate intracellular bacteria are devoid of genetic parasites. Instead, we propose a more intricate picture—one in which differences in the modes of transmission of these bacteria partly predict distinct genomic outcomes for mobile DNA in obligate intracellular species.

22.3.2   Composition of mobile DNA

Compares the number of genes that have mobile-DNA functions in obligate versus facultative intracellular bacteria. The bacteria represented in comprise most species that have well defined intracellular lifestyles and genomes that have been completely annotated in the Comprehensive Microbial Resource v15.2 of The Institute for Genomic Research (TIGR). To generate this database, TIGR carries out an automated annotation of completed microbial genomes and classifies genes into 19 functional-role categories, which include a mobile-DNA category that specifies prophage, transposable element and plasmid genes.

Other types of horizontally acquired or mobile DNA such as pathogenicity islands, integrons and conjugative transposons are not catalogued in this analysis.

First, the amount of mobile DNA in the genomes of obligate intracellular species spans almost the same range as the genomes of facultative intracellular species. However, facultative intracellular bacteria contain an average of four-fold more mobile DNA than obligate intracellular bacteria ($p = 0.0015$, Mann–Whitney U-test). This finding is consistent with predictions that facultative intracellular bacteria have mobile-DNA compositions that are more similar to free-living than to obligate species. It also dispels the assumption that obligate intracellular bacteria lack mobile genetic elements or their remnants — the genomes of at least half of the obligate species analysed contained some mobile DNA. However, with the exception of *Wolbachia pipientis* wMel, in obligate intracellular species, mobile DNA comprises less than 2% of the total genome — a level that is similar to the lower end of the range found in facultative intracellular species.

The second finding answers a basic question that has emerged from genomic analyses — what type of mobile genetic element is most common in intracellular bacteria? Transposable elements constitute the largest portion of mobile DNA. The proportion of plasmid genes per genome is consistently small, and the proportion of prophage genes per genome is intermediate to that of plasmid and transposable-element genes. Transposable elements might predominate in bacterial genomes because they often do not require site specificity for insertion and can integrate into a genome that already has a copy of the same transposable element. By contrast, genome insertion by phages generally shows site specificity and confers immunity to multiple infections. It should also be noted that phages serve as vectors that shuttle other mobile elements, such as transposable elements, into a host genome. However, the difference between the amount of transposable-element and prophage-related genes found in intracellular bacteria is nonetheless striking, as a transposable element typically carries a single gene (encoding a transposase or reverse transcriptase/maturase), whereas a prophage genome consists of tens of genes.

Third, this analysis provides a first glimpse of the impact of lifestyle differences on the mobile-DNA composition of obligate intracellular species. The genera that have mobile-DNA genes are strictly pathogenic (*Coxiella*, *Chlamydia*, *Chlamydophila*, *Phytoplasma* and *Rickettsia*) or parasitic (*Wolbachia*) genera that undergo at least some horizontal transmission. By contrast, those obligate species that lack mobile genetic regions include all of the dietary mutualists of insects that are vertically transmitted (*Wigglesworthia*, *Blochmannia* and three strains of *Buchnera*). This indicates that transmission differences among obligate intracellular species might shape genome plasticity. In particular, the evolutionary processes that operate in the small population sizes of these strictly maternally inherited species (for example, GENETIC DRIFT) might accelerate the loss of mobile DNA. By contrast, the more permissive lifestyles of host-switching pathogens and parasites might augment their contact with novel gene pools and the uptake of foreign DNA. One exception could be the unpublished report of a high mobile-DNA content in the *Sitophilus oryzae* primary endosymbiont (SOPE), which is a recently derived, γ-proteobacterial mutualist of *Sitophilus* grain weevils. Maternally transmitted bacteria that have recently adopted an intracellular lifestyle might have mobile-DNA contents that are more similar to their free-living or facultative intra cellular relatives, because not enough time has elapsed to streamline the genome by deletion of mobile DNA.

22.3.3   Correlation of mobile DNA with genome size

The relationship between the total number of genes and the proportion of mobile DNA in a genome reflects the balance between the inflow and outflow of mobile DNA. Whereas gene-gain events depend on rates of horizontal gene transfer and gene duplication, gene-loss events depend on rates of gene inactivation and deletion. If mobile-DNA loss is enhanced in the genomes of long-term obligate intracellular species or if mobile-DNA gain is enhanced in the genomes of facultative intracellular species, the fraction of mobile-DNA genes in the obligate species will be reduced compared with that of facultative or free-living species. Alternatively, if the loss and gain of mobile DNA is random, the proportion of mobile-DNA genes should remain constant across species. Genome analysis shows that the relative mobile-DNA composition of intracellular bacteria increases with total gene number. Therefore, although we now know that obligate intracellular species do have mobile genetic elements, the smaller genomes of obligate intracellular species show preferential deletion of these mobile elements.

The conventional explanation for the reduction of mobile DNA in obligate intracellular bacteria is that these bacteria have a general mutational bias for deletions and, therefore, the rate of gene loss in such species exceeds the rate of new gene acquisitions by horizontal gene transfer or gene duplications. This bias is thought to result in the relatively small genomes of prokaryotes and the close packing of genes in bacterial genomes. Indeed, a comparative analysis of pseudogene evolution across a wide range of prokaryotes shows that deletions are far more frequent than insertions. Elevated deletion rates in bacteria might even be due to deterministic evolutionary forces such as selection against the continuous influx of dangerous genetic parasites.

Therefore, an inherent mutational bias for deletions in prokaryotes, coupled with no or relatively limited inflow of new mobile elements in endosymbionts, could accelerate mobile-DNA reduction in obligate intracellular species.

22.3.4 Plasmids

Plasmids are extrachromosomal elements that move horizontally between bacterial cells in at least four ways: self-directed transmission, in which the plasmid encodes the conjugative machinery for host-cell fusion; mobilizable methods, in which one plasmid parasitises the self-directed transmission of another plasmid; transduction, in which a plasmid gets packaged in a phage particle; and transformation, in which cell lysis releases plasmid elements from the host bacterium. Plasmids also move vertically by transmission through dividing host cells. Similar to phages, plasmids might have an important role in microbial evolution by acting as natural gene vectors.

Among intracellular bacteria, instances of plasmid genes inserted into the host chromosome are uncommon. Only two of the obligate intracellular bacterial genomes have genes with plasmid functions. Of these species, the maximum number of plasmid genes per genome is five (*Coxiella burnetti*), and the average number of plasmid genes in obligate and facultative intracellular bacteria is approximately one. Plasmid genes therefore do not significantly affect the genetic architecture of intracellular bacterial genomes. However, this does not preclude their importance as natural gene vectors of foreign DNA that might be advantageous or deleterious to the recipient host genome.

Among obligate intracellular prokaryotes, at least six different genera harbour extrachromosomal plasmids, including the insect mutualists *Buchnera*, *Wigglesworthia* and *Sodalis* and the obligate intracellular parasites *Chlamydia*, *Chlamydophila* and *Phytoplasma*. The two plasmids in *Buchnera* are vertically transmitted and carry amino-acid biosynthesis genes that aid the main role of *Buchnera* symbionts in aphids — the provision of amino acids that are deficient in the insect phloem-sap diet. However, not all *Buchnera* strains carry plasmids, and the *leuABCD* plasmid might experience only rare horizontal transmission and genetic exchange with the *Buchnera* host chromosome. Much less is known about the plasmids that have been isolated from *Wigglesworthia glossinidia* and *Sodalis glossinidius*. The former has a single 5.3-kb plasmid called pWig1, and the latter has a 134-kb plasmid and possibly a 10-kb plasmid. Also, members of the diverse genus of *Phytoplasma* have at least 12 plasmids of unknown function. All are less than 11 kb in size, share varying amounts of homology and structure, and frequent rearrangements might affect their size and structure. Plasmid pOYM notably encodes a unique replication protein that has domains that are related to both prokaryotic plasmids and eukaryotic viruses.

22.3.5 Bacteriophages

Bacteriophages are viruses of prokaryotes and are among the most abundant biological entities in the biosphere. They are one of the most effective vectors that convey foreign DNA into recipient bacteria and can cause significant amounts of genome diversification. Virulent phages always lyse their hosts after invasion, whereas temperate phages can integrate their DNA into the bacterial host chromosome as a prophage, so that the DNA is passively inherited. On prophage integration, associated DNA elements, such as pathogenicity islands or other mobile genetic elements, are also transferred into the host chromosome. Prophages consequently account for large fractions of horizontally acquired DNA in bacterial genomes. Intact, complete phages have been purified, isolated and sequenced from two groups of obligate intracellular bacteria, *Wolbachia* and *Chlamydiaceae*. A third case of phage infection occurs in the secondary endosymbionts (γ-proteobacteria) of aphids, which have not been firmly characterised as obligate or facultative intracellular bacteria.

Wolbachia *of arthropods*. *Wolbachia* are a genus of cytoplasmically transmitted α-proteobacteria that infect millions of arthropod species and many filarial nematodes, and which are phylogenetically related to emerging human pathogens in the Anaplasmataceae. Their roles in inflammatory-mediated FILARIAL DISEASE and their unusually high levels of genomic and phenotypic plasticity have attracted recent interest. *Wolbachia* are most well known for inducing several selfish forms of reproductive parasitism in arthropods, which might affect key evolutionary processes, including speciation, sex determination and sexual selection. Despite early microscopy studies that reported phage particles in *Wolbachia* of the mosquito *Culex pipiens*, insect endosymbionts in general were previously presumed to be refractory to mobile genetic parasites. However, after the isolation and sequencing of prophage WO (for *Wolbachia*) in 2000, the analysis of molecular markers and phage-filtration techniques established the connection between the virus particles of *Wolbachia* and prophage WO. The genome sequence of *Wolbachia* strain wMel from *Drosophila melanogaster* reveals two divergent prophage WO families, WO-A and WO-B. Molecular evolution analyses indicate that family WO-B groups into three CLADES and distribution surveys indicate that WO-B homologues occur in at least 89% of two main lineages of *Wolbachia* that infect arthropods. Notably, WO-B is the only bacteriophage found in insect endosymbionts that is known to elicit gene expression, carry transposable elements, recombine at fast rates and laterally transfer between bacteria. Owing to its high prevalence in *Wolbachia*, WO-B might be one of the most abundant viruses in all obligate intracellular bacteria, and it could be an important source of genomic flux in the evolutionarily important *Wolbachia*.

The *Wolbachia* wMel genome sequence also contains a small pyocin-like element (prophage Py) that contains nine genes. Pyocin elements encode bacterial products that morphologically resemble bacteriophage tails and that often have bactericidal activities towards bacterial strains and species that are closely related to the producer. All of the *Wolbachia* *w*Mel phage elements have a low GC content, similar to the host chromosome, which is indicative of a long association with *Wolbachia* and/or other intracellular bacteria. Because these elements do not exist in *Rickettsia* relatives, we infer that they entered the *Wolbachia* system from an intracellular species that is not closely related to the Anaplasmataceae family of α-proteobacteria. However, ongoing lateral exchange of bacteriophage WO-B between different strains of *Wolbachia* on co-infection of cells does seem to take place and these cases represent the first clear findings of recent lateral transfer in obligate intracellular bacteria.

Chlamydiaceae: The family Chlamydiaceae is an abundant group of obligate intracellular bacteria composed of two separate genera, *Chlamydia* and *Chlamydophila*, which contain three and six species, respectively. *Chlamydia trachomatis* and *Chlamydophila pneumoniae* are human pathogens. All Chlamydiaceae exist in two alternating developmental forms, which include a metabolically inert extra cellular form, termed the elementary body, and an intracellular replicating cell, termed the reticulated body. Virus particles that infect the reticulated body were first observed in the avian strain *Chlamydophila psittaci* and at least five single-stranded DNA (ssDNA) Microviridae bacteriophages are now known to infect the *Chlamydophila* genus. All five phages constitute a divergent subfamily of the well known coliphage φX174. In total, four of the phages, φCPG1, φCPAR39, Chp2, and Chp3, are closely related, with genome identities greater than 90% and these are divergent from the fifth phage, Chp1. The genomes of all five bacteriophages encode at least three viral structural proteins and a few additional genes. Phage infection typically reduces bacterial viability in the laboratory, but further work remains to determine how phage infection generally impacts chlamydial disease, evolution and genetics. Recent horizontal acquisition of phage φCPAR39 is inferred, based on its sporadic distribution across nearly identical *C. pneumoniae* host strains. The five ssDNA phages of the *Chlamydophila* genus are unrelated to the double-stranded-DNA phage WO-B, and therefore represent a completely separate event of phage evolution in obligate intracellular bacteria.

22.3.6   Transposable elements

Transposable elements are mobile genetic elements that can move (transpose) from one site in the genome to a second site, or from one DNA molecule (that is, an infecting phage genome or a plasmid) to a second DNA molecule (the bacterial chromosome). They are the most abundant type of mobile genetic element in the genomes of intracellular bacteria. The number and distribution of Transposase genes in various intracellular bacteria analysed from the TIGR database. Five out of thirteen obligate intracellular species and all thirteen of the facultative intracellular species harbour transposable elements. Among these species, the average number of transposable elements varies, with obligate intracellular bacteria and facultative intracellular bacteria harbouring an average of 29.2 and 131.9 transposase genes, respectively. There is also a report of an Insertion sequence (IS) element in the primary Endosymbiont Sope of the *Sitophilus zeamais* grain weevil, the complete genome of which has not been sequenced.

Two classes of transposable elements are found in the genomes of obligate intracellular bacteria. The most 'genetically destructive' class is the DNA transposable element. These elements are identified by the presence of transposase-encoding genes that are members of transposase families that have previously been found in free-living bacterial species. For instance, *Wolbachia* *w*Mel contains transposase genes from the widely dispersed IS*3*, IS*4*, IS*5*, IS*110* and IS*256* families. DNA transposable elements from these families transpose by conservative mechanisms described. The impact of these transposition events includes possible inefficient repair of donor DNA molecules and the disruptive effects of the transposable element insertions. Preliminary data also indicate the presence of previously undetected IS elements (IS*21*, IS*3* and IS*481*) in the genomes of *Buchnera* sp. APS, *Chlamydia muridarum* and *C. trachomatis*.

The second group of transposable elements comprises the retrotransposable elements that move through an RNA intermediate. The presence of these elements is indicated by the discovery of genes that encode reverse transcriptases (RT) in *Wolbachia* *w*Mel. There are fourteen RT-like genes, many of which are identical or nearly identical. Of particular interest are three intact RT genes that contain MATURASE DOMAINS (WD0693, WD0995 and WD1138). A maturase domain in the RT genes is a strong genetic signature of a MOBILE GROUP II INTRON. Mobile introns should have no impact on donor DNA sequences and little impact on the integrity of target genes if they are inserted in the correct orientation, as splicing will remove the intron sequences from the target-gene mRNAs. Why there is a surplus and diversity of RT ORFs in *Wolbachia* remains a question for future study, but their presence in *Wolbachia* and absence from related *Rickettsia* genomes implies that they originated through horizontal-RT-gene transfer events.

22.3.7   Disrupted Mobile-DNA Genes

Eight of the thirteen obligate intracellular and all of the thirteen facultative intracellular species analysed in this review have genes that are annotated with mobile-DNA functions. Three patterns are evident with respect to the proportion of these mobile-DNA genes that are inactivated owing to mutations. First, the total proportion of disrupted mobile-DNA genes in the obligate

(31.4%, 59/188) and facultative (0.9%, 22/2,414) intracellular species indicates an asymmetry in the coding capacity of mobile DNA between these two classes of bacteria. This asymmetry is consistent with the preferential degenerative evolution of mobile DNA in obligate intracellular species. A closer look at the type of inactivated genes indicates a tendency to inactivate transposable element genes across both classes of bacteria. In the three obligate intracellular species with inactivated mobile DNA, including *Wolbachia w*Mel, *C. burnetti* and *Chlamydophila caviae*, almost all of the disrupted genes encode presumed transposases or RTs. Similarly, in the three facultative intracellular species, including *Mycobacterium smegmatis*, *Mycobacterium tuberculosis* and *Listeria monocytogenes* 4b, almost all of the disrupted genes encode transposases.

Therefore, whereas obligate intracellular species are more likely to have disrupted mobile-DNA genes, perhaps owing to accelerated rates of mutation and gene inactivation, there seems to be a common selective force in bacteria that specifically drives inactivation of DNA transposable elements. A high fraction of nonsite-specific DNA-transposition events would generate lethal mutations, ultimately posing a greater selective pressure against the presence of active transposable elements in a genome. Furthermore, excision of a DNA transposable element requires double-strand-break repair of the transposable element donor chromosome, and this seems to be inefficient. However, this does not explain the apparent high levels of RT-gene inactivation, as retrotransposition should have less damaging consequences. Instead, it is possible that the error-prone nature of RT activity results in an elevated mutation frequency for RT genes. By contrast, phages often show site specificity in invading a host genome and might pose less of a fitness cost to their host bacterium if these sites are non-essential. Moreover, excision of prophages typically is efficient and self-repairing. Therefore, DNA-transposable-element inactivation might reflect deterministic forces to specifically eliminate harmful transposable elements from the genome. Mobile DNA and the 'intracellular arena' The presence of mobile DNA in obligate intra cellular bacteria raises two crucial questions: first, what is its origin, and second, how does it spread in a host population of obligate intracellular bacteria? The first question can be answered using phylogenomic studies. Assuming that the mobile genetic elements in intra cellular bacteria were originally derived from orthologous elements in free-living species and that enough microbial genomes will be sequenced to trace the historical pedigree of mobile-DNA donors and recipients, then phylogenetic analysis should piece together the evolutionary history of mobile DNA in obligate intracellular bacteria.

The amount of genome-sequence information that is currently available paints a cloudy picture of the origins of mobile DNA in obligate intracellular species. Although putative orthologues of transposable elements and bacteriophages can be identified among free living and intracellular bacteria, we are unable to discern between the original and most recent donor species of mobile DNA. Two interesting cases of transposable element acquisitions in the *Wolbachia w*Mel genome indicate that the bacteriophage WO-B might be a vector for introducing transposable elements into *Wolbachia*. An IS*110*-like transposase sequence from bacteriophage WO-B from *Wolbachia w*CauB is present in multiple (and often degenerate) copies throughout the *w*Mel genome (both within and distant from prophage regions) and is sometimes flanked by additional transposase sequences. There is also an intact IS*50*-like sequence in the WO-B prophage genome of *Wolbachia w*Tai. IS*50*-transposase homologues are rare, but occur in widely divergent bacteria, including *Wolbachia w*Mel. The most recent donors of the bacteriophages themselves are probably other intra cellular bacteria, as the shared AT nucleotide biases of intracellular bacteria with mobile-element sequences indicate a long association in species with an intra cellular lifestyle.

There are several possible answers to the second question — how mobile DNA spreads among the obligate intracellular bacteria. Mobile DNA might confer a fitness benefit to the host and spread rapidly, but, so far, there is no clear evidence that transposable elements or bacteriophages carry genes that are adaptive for their obligate intracellular hosts. It is more probable that the mobile elements are neutral or even deleterious to the bacterial host.

The 'intracellular arena' hypothesis posits that genetic material can move in and out of communities of obligate intracellular bacteria that co-infect the same intracellular host environment. Therefore, one could view the eukaryotic host cell as a consortium of co-infecting intracellular bacteria that span different genotypes, species, genera and major orders of bacteria. Such an arena of interacting micro-organisms could provide an escape from 'genetic confinement' for these specialised micro-organisms and a window of opportunity for recombination and horizontal exchange of genomic elements.

This new hypothesis is motivated by reports of diverse bacteria that can co-infect the same host and molecular evolution studies that have examined chromosomal recombination and lateral exchange of phage DNA. The α-proteobacteria *Wolbachia* often co-infect hosts with many other intracellular bacteria, including other *Wolbachia* strains, Anaplasmataceae relatives, γ-proteobacteria and the Bacteroidetes parasite, *Candidatus Cardinium hertigii*. The secondary endosymbionts of pea aphids horizontally transfer and co-infect aphids with primary endosymbionts and other strains of secondary endosymbionts at high frequencies. Furthermore, the arthropodborne pathogens *Rickettsia* and *Anaplasma* co-infect the same tick hosts, and different strains of the plant-pathogen *Phytoplasma* are also known to establish mixed infections in their plant hosts. Also, *C. pneumoniae* can establish mixed infections in human hosts with *Mycoplasma pneumoniae* and *Streptococcus pneumoniae*. If there is transfer of mobile or chromosomal DNA among these microbial communities, then the view that obligate intracellular bacteria are devoid of genetic exchange might be overly simplistic. So far, there is evidence in *Wolbachia* that supports the intracellular arena hypothesis. In three cases of insect species infected with two divergent strains of *Wolbachia*, recent lateral transfer of phage WO-B between the co-infecting strains was inferred based on comparative sequence analyses of a capsid-protein gene. Results indicate that this evolutionarily recent exchange of phage DNA either occurred through recombination between prophage DNA sequences, horizontal transfer of complete or partial phage genomes or recombination between DNA from prophage and phage particles, as has been recently proposed in certain systems. Determining the extent of horizontal DNA transfer associated with phage elements will be an interesting area of future research. Comparative analyses across sets of *Wolbachia* gene sequences specify that recombination frequently occurs among closely related strains and possibly between divergent groups in the gene encoding the *Wolbachia* surface protein Wsp. The extent to which genetic exchange in the intracellular arena influences the intracellular microbial community and the spread of mobile genetic elements in bacterial communities should be a topic of future study.

Interestingly, the recently sequenced genome of a *Wolbachia* strain that infects the pathogenic nematode *Brugia malayi* was found to have far fewer mobile-DNA genes than the genome of *Wolbachia* wMel. This streamlined version of the *Wolbachia* genome is not surprising when one considers its strict vertical transmission and obligate mutualistic lifestyle. All filarial nematodes are only infected by single strains of *Wolbachia* that are typically reduced in genome size compared with the *Wolbachia* that infect arthropods. The disparity in mobile-DNA content across a monophyletic clade of bacteria with varied transmission routes and host ranges clearly highlights the effects of lifestyle differences on mobile-DNA acquisition and invasion.

22.4 Mobile DNA and evolution in the 21st century

Just as our knowledge of mobile DNA has introduced new ways of thinking about hereditary change, the results of sequence analysis have documented several types of genome alterations at key places in evolutionary history, alterations which are notable because they happened within a single generation and affected multiple cellular and organismal characters at the same time: horizontal transfers of large DNA segments, cell fusions and symbioses, and whole genome doublings (WGDs). These rapid multi-character changes are fundamentally different from the slowly accumulating small random variations postulated in Darwinian and neo-Darwinian theory.

Cell mergers and WGDs are the kinds of events that activate mobile DNA and genome restructuring. In order to fully integrate the genomic findings with our knowledge of mobile DNA, we have to make use of information about the molecular regulation of mobile DNA activities as well as McClintock's view that cells respond to signs of danger, frequently restructuring their genomes as part of the response. This regulatory/cognitive view of genome restructuring helps us to formulate reasonable hypotheses about two unresolved questions in evolutionary theory: (i) the connections between evolutionary change and ecological disruption and (ii) the origins of complex adaptive novelties at moments of macroevolutionary change.

We can summarise the change from the simple linear view of the Central Dogma to todays complex systemsbased picture of cell informatics by writing out molecular information transfer events as sets of Boolean propositions (adapted from):

Crick's central dogma of molecular biology:

1. DNA → 2X DNA.
2. DNA → RNA → protein → phenotype.

Contemporary picture of molecular information transfers:

1. DNA + 0 → 0.
2. DNA + protein + ncRNA → chromatin.
3. Chromatin + protein + ncRNA → DNA replication, chromatin maintenance/reconstitution.
4. Protein + RNA + lipids + small molecules → signal transduction.
5. Chromatin + protein + signals → RNA (primary transcript).
6. RNA + protein + ncRNA → RNA (processed transcript).
7. RNA + protein + ncRNA → protein (primary translation product).
8. Protein + nucleotides + Ac-CoA + SAM + sugars + lipids → processed and decorated protein.
9. DNA + protein → new DNA sequence (mutator polymerases).
10. Signals + chromatin + protein → new DNA structure (DNA rearrangements subject to stimuli).
11. RNA + protein + chromatin → new DNA structure (retrotransposition, retroduction, retrohoming).
12. Signals + chromatin + proteins + ncRNA + lipids → nuclear/nucleoid localisation SUMMARY: DNA + protein + ncRNA + signals + other molecules ⟷ Genome structure and phenotype.

A helpful analogy for the role of the genome in cellular informatics is as a RW memory system. This is a fundamentally different idea from the conventional

20th century view of the genome as a read-only memory (ROM) subject to accidental change. DNA is a multivalent storage medium capable of holding information in nucleotide sequences, chemical modifications, and nucleoprotein complexes. In thinking about how the cell writes information back onto the genome, we can discriminate roughly three different time scales:

1. Within the cell cycle, where the formation and dissolution of transient nucleoprotein complexes predominate.

2. Over several cell cycles, where heritable chromatin configurations can be passed on and then erased or re-imprinted.

3. Over evolutionary time, where sequence variation and genome restructuring play major roles in the emergence of novel characters and adaptive functions.

22.4.1   Mobile elements and regulatory evolution

Transcription signals

The appearance of a novel coding capacity at a genetic locus frequently results from changes in *cis*-acting regulatory and processing signals without any change in exon content. Mobile DNA has long been known to play a role in this kind of regulatory change. The phenotypes of the first bacterial mutations known to be IS (insertion sequence) elements resulted either from the acquisition of transcriptional stop signals or from the creation of novel transcriptional start sites. In eukaryotes, mutations activating transcription most commonly resulted from the insertion of enhancer elements in LTR (long terminal repeat) retro-elements. In the case of one apoptosis regulator protein, genome comparison shows that orthologous coding regions in primates and rodents acquired their parallel transcription signals from independent LTR retrotransposon insertions. Sequences of Mu element insertions in maize can alter both the initiation and termination sites for transcription. Examination of the human genome has uncovered over 100 cases where Alu elements provided polyA addition signals at the 3′ end of expressed sequences. The role of mobile elements in the evolution of transcriptional regulatory sites has been extensively documented from genomic data since the 1990s. Many of these cases display the kind of taxonomic specificity predicted by the phylogenetic distribution of transposons and retrotransposons.

Splicing signals

It has been over two decades since Wessler and colleagues discovered the splicing of Ds insertions in maize. Not only does Ds behave as a mobile intron; it also confers alternative splicing. The same is true of maize retrotransposons. The potential of a single genomic change to encode multiple novel products has been documented in broad beans, where insertion of a CACTA family transposon carries out exon shuffling and provides sites for alternative splicing. Recent studies in the human genome are beginning to clarify the requirements for generating novel splicing patterns by mobile element inserts.

Chromatin signals

The insertion of a mobile element has a profound effect on local chromatin configuration. Since a major regulatory mechanism for controlling the activity of mobile elements is incorporation into silenced chromatin, individual or clustered elements serve as nucleation sites for heterochromatin domains. Some elements, like gypsy in Drosophila, carry chromatin insulator determinants that are major contributors to their influence on genome expression. In certain cases, like the FWA and MEDEA loci in Arabidopsis, imprinted expression reflects the action of RNAi machinery on sequences derived from a mobile element. Recent studies of imprinted loci in Arabidopsis seeds indicate that mobile elements provided many of the recognition sequences for epigenetic control. The connection between mobile elements and chromatin signals is less well-documented in mammals. Nonetheless, there is intriguing evidence that retrotransposons were critical to the origin of an epigenetic control regime necessary for the emergence of mammals in evolution. Knockout experiments in mice show that imprinted loci derived from the Ty3/gypsy retrotransposon family are essential to placental development. These observations suggest that functional exaptation of retrotransposon coding sequences and signals mediating their epigenetic control played a role in the evolution of the placenta, a major developmental invention.

Regulatory RNAs

We are currently learning how much regulation occurs through the action of small RNA molecules. The examination of plant genome sequences has established important links of many small RNAs to DNA transposons (miniature inverted-repeat transposable elements - MITEs) and led to the suggestion that si- and miRNA regulation evolved from mobile element controls. The rice Pack-MULEs are also associated with small RNA coding sequences. In the human genome, 55 functionally characterised and 85 uncharacterised miRNAs arose from transposons and retrotransposons. Comparison with the mouse genome indicates that miRNAs matching L2 LINE and MIR SINE elements are ancient and conserved, while those matching L1 LINE and DNA elements are primate-specific. As expected from the taxonomic distribution of SINE elements, the Alu-derived miRNAs are also primate-specific. Alu element recombination also appears to have played a role in the expansion of primate miRNA coding arrays.

A similar conclusion about the role of mobile elements in the generation of taxonomically-specific miRNAs arose from analysis of marsupial genomes.

24.4.2   Intercellular horizontal DNA transfer

Molecular genetics began with the study of intercellular horizontal DNA transfer. The first demonstration of the genetic capacity of DNA molecules involved pneumococcal transformation and bacterial genetics developed on the basis of cells' capacities to transfer genome segments by transformation, conjugation or viral transduction. Studies of temperate bacteriophages and antibiotic resistance made us appreciate the multiple molecular mechanisms cells have to incorporate newly acquired DNA independently of extensive sequence homology. From countless experiments, we now have overwhelming evidence for horizontal DNA transfer between species and between the three kingdoms of living cells (Table 22.1).

Table 22.1: Modes of intercellular and interkingdom DNA transfer.

Horizontal transfer mode	Documented transfers
Uptake of environmental DNA	Bacteria – bacteria
	Plant – bacteria
	Plastid transfection
	Mammalian cell transfection and lipofection
Conjugal transfer	Bacteria – bacteria
	Bacteria – yeast
	Bacteria – plant
Viral transduction and gene transfer agents	Bacteria – bacteria
	Bacteria – plant
	Animal cell – animal cell
	Animal cell – virus
Parasitic or endosymbiotic association	Plant – fern
	Plant – plant
	Bacteria – invertebrate
Undetermined mechanism	Archaea – bacteria

Horizontal transfer can be a major driver of evolutionary novelty because it permits the acquisition of DNA encoding complex traits in a single event. The genomic data is overwhelming in documenting the fundamental importance of horizontal transfer in the evolution of bacterial and archaeal genomes. Prokaryotic genomes contain plasmids and genomic islands encoding multi-component adaptive characters that range from microbicide resistance, virulence and symbiosis to metabolism and magnetotaxis. This has led to a scheme of bacterial and archaeal evolution which has a reticular rather than a branching structure. The possibility that different genome components could display different phylogenies due to horizontal transfer was quite literally inconceivable to Darwin and his mid-20th century neo-Darwinian successors.

Although we have long been familiar with the prokaryotic story, there is rapidly growing evidence for intercellular and interkingdom horizontal transfer events in the evolutionary history of eukaryotic genomes. The data include phylogenetically dispersed coding sequences and mobile elements, as well as the incorporation of genomic segments from prokaryotic and eukaryotic endosymbionts and parasites. There is also evidence of host-to-parasite transfer. In certain microbially diverse ecosystems, such as the rumen, frequent prokaryote to eukaryote transfer occurs. In plants, but not animals, there is extensive horizontal transfer of mitochrondrial DNA. Similar transfer is very rarely seen in the plastids, which may be explained by the fact that the mitochondria have a DNA uptake system not found in chloroplasts. The functional consequences of horizontal transfer into eukaryotes range from the acquisition of single bio-chemical activities to major restructuring of metabolism to integrating multiple functions needed to occupy new ecologies, as illustrated by fungal pathogens, the anaerobic human parasites Entamoeba histolytica and Trichomonas vaginalis and plant parasitic nematodes.

24.4.3   Cell fusions and intracellular DNA transfer at key junctures in eukaryotic evolution

One of the early accomplishments of nucleic acid sequencing was to confirm the endosymbiotic origin of mitochondria and plastids. Combined with evidence that the mitochondrion is an ancestral character for all eukaryotes, this confirmation places cell fusion events at the root of eukaryotic evolution. For photosynthetic eukaryotes, the original cyanobacterial fusion that generated the ancestral plastid has been followed by a series of secondary symbioses between various eukaryotic lineages and either red or green algae. The most 'basal' photosynthetic lineage appears to be the glaucophytes, because their plastids retain bacterial peptidoglycans. Through evidence of cell fusions and endosymbiosis, genome sequencing has introduced another major process of rapid and multicharacter change into the established evolutionary record. Lacking knowledge of cell biology, such a mechanism of variation was not considered by Darwin and has been largely ignored by his neo-Darwinian followers.

As the following descriptions of various endosymbioses show, DNA mobility between distinct genome compartments was a major feature of adjustment to cell fusion events. Sequence evidence indicates that all the cell fusions in eukaryotic lineages were followed by massive episodes of intracellular horizontal

DNA transfer between the organelle and nuclear genomes. That is why the majority of organelle proteins are encoded by the nuclear genome. Moreover, these organelle genomes are remarkably dynamic in their evolution. Mitochondria display a great range of genome size (~6 kb to ~480 kb), and a number of them have strikingly elaborate DNA structures (for example, multiple linear molecules, interlocked circles) and/or modes of expression. There are anaerobic eukaryotes that have lost the oxidative functions of mitochondria, but most of them retain related organelles labelled hydrogenosomes or mitosomes.

The history of plastids, descended from cyanobacteria, is somewhat different from that of mitochondria, descended from alpha-protobacteria. In higher plants and photosynthetic algae, the chloroplast genome is relatively stable and falls within a relatively narrow size range of 120 kb–160 kb. In heterotrophic or parasitic species that have lost photosynthesis, the plastid genome is reduced but still retained at sizes greater than 34 kb. In the apicomplexan parasites, plastid genomes are known to have undergone extensive structural rearrangements.

Non-photosynthetic chloroplast derivatives appear to retain residual functions, such as encoding tRNAs that may be used by mitochondria, activities involved in the biosynthesis of amino acids, fatty acids, isoprenoids, heme, pigments and enzymes for detoxifying oxidative radicals.

In cells of organisms arising from secondary symbioses with red algae (cryptomonads) or green algae (chlorarachniophytes), there are actually four distinct genome compartments: nucleus, mitochondrion, plastid and nucleomorph (the descendant of the algal nucleus). The plastid and nucleomorph compartments are surrounded by four, rather than two, membranes which, presumably, is a reflection of their origins by phagocytosis. The two sequenced nucleomorph genomes are 551 kb (Guillardia theta, cryptomonad) and 373 kb (Bigelowiella natans, chlorarachniophyte), each containing three chromosomes with telomeres. These genomes encode their own 18S eukaryotic ribosomal RNA, other RNAs and proteins (465 and 293, respectively). The nuclear genomes of both species contain coding sequences of red- or green-algal origin, indicating extensive intracellular horizontal transfer.

In addition to the remarkable multi-genome cells just described, there are cases of tertiary symbioses in the dinoflagellates, which have fused with green algae, haptophytes, diatoms and cryptomonads. It appears, from the analysis of the origins of nuclear coding sequences for plastid-targeted proteins, that dinoflagellates and other chromalveolates have retained an ability to phagocytose other cells and recruit fragments of their genomes, but that the capacity was lost in the photosynthetic lineages leading to green algae, plants and red algae.

24.4.4 Whole genome doublings at key places in eukaryotic evolution

Genome sequencing has made it clear how important the amplification and modification of various genome components has been. Of particular interest has been the formation of families of coding elements for homologous proteins within genomes. Both prokaryote and eukaryote species encode characteristic protein families, which are important guides to the functions those species need to thrive in their particular ecological niches. As complete genome sequences accumulated, it became apparent that not only the genetic loci encoding individual proteins had amplified; large chromosome regions had also undergone duplication processes. These 'syntenic' regions carry genetic loci in the same order and orientation. By comparing related taxa, it has been possible to discern phylogenic branches that have inherited two copies of multiple ancestral segments. These segments are now understood to be the remnants of WGD events at the base of the branch.

Genome doublings have been documented in yeasts, ciliated protozoa and plants. There is even evidence of a genome tripling at the base of the angiosperm radiation. In animals, the most important WGD events have been found at the base of the vertebrate lineage, where two successive events gave rise first to all vertebrates and then to jawed vertebrates. Later in vertebrate evolution, there was another WGD event at the origin of teleost fish. Characteristic of transitions marked by WGD events are the rapid formation of a cluster of related species, as in Paramecium or the appearance of major innovations, as with the vertebrate skeleton and jaw. WGD is yet another evolutionary process outside the Darwinist perspective that occurs suddenly (that is, within a single generation) and simultaneously affects multiple phenotypic characters. It is especially significant to note that a genome doubling means that the dispersed coding elements for complex circuits are duplicated and the two duplicate circuits can then undergo independent modifications as distinct entities.

There is an important connection between WGD and synthetic speciation. It is possible to generate new species of plants by interspecific hybridisation and genome doubling. Fertile hybrids tend to have tetraploid genomes. Genome doubling helps maintain stability through meiosis because each chromosome in the hybrid has a homologous partner for pairing and crossing over. There is also evidence that genome doubling helps maintain normal transcription patterns. The genome duplication events may occur either during gametogenesis or after fertilisation, but in plants the most common process involves diploid gametes. The incidence of spontaneous genome doubling is surprisingly high, reaching 1% of all fertilisations in mice. It is of great theoretical significance that synthetic speciation takes place rapidly after hybridisation rather than slowly following repeated selections, as predicted by conventional theory. The evolutionary importance of interspecific hybridisation in promoting evolutionary change has been appreciated since a time predating the molecular genetics revolution. Although most synthetic and observational work has been done with plants, there are reports of contemporary natural hybridisation involving animals. The animal cases include Darwin's finches in the Galapagos Islands, long taken as a paradigm of gradualist evolution. The finch case is especially instructive because hybridisation leads to abrupt, unpredictable changes in beak shape. Responses of mobile DNA systems to infection, hybridisation and genome duplications The genomic evidence showing that cell fusions and WGD have occurred at key junctures in eukaryotic evolutionary phylogenies leads to the question of what effect such events (plus the related process of interspecific hybridisation) have on mobile DNA and natural genetic engineering functions. The answer is that all these processes are major triggers of genomic instability and restructuring, with microbial infection serving as a proxy for cell fusions. The data on hybridisation responses are more extensive in plants, but we have enough cases in animals to be confident that the answer there is equally valid. Moreover, we know of many cases of hybrid dysgenesis in animals, where activation of mobile elements and widespread genomic changes results from inter-population mating. In at least one intriguing plant case, interspecific mating has triggered genomic instability with formation of a zygote containing only one of the parental genomes. The rapid natural genetic engineering response to genome doubling reflects a tendency to return to the normal diploid state. This poorly understood process of diploidisation involves chromosome loss, deletions and chromosome rearrangements. The chief mechanistic basis for activation of natural genetic engineering in response to hybridisation and genome doubling appears to be changes in chromatin organisation and in epigenetic modifications of the DNA that normally inhibit activity of mobile elements.

24.4.5 Evolutionary advantages of searching genome space by natural genetic engineering

One of the traditional objections to Darwinian gradualism has been that it is too slow and indeterminate a process to account for natural adaptations, even allowing for long periods of random mutation and selection. A successful random walk through the virtually infinite dimensions of possible genome configurations simply has too low a probability of success. Is there a more efficient way for cells to search 'genome space' and increase their probability of hitting upon useful new DNA structures? There is, and the underlying molecular mechanisms utilise the demonstrated capabilities of mobile DNA and other natural genetic engineering systems.

Perhaps the most important aspect of evolutionary change by natural genetic engineering is that it employs a combinatorial search process based upon DNA modules that already possess functionality. The evolutionary reuse of functional components has been recognised for many years, but it is only with genome sequencing that we have come to appreciate how fundamental and virtually ubiquitous such reuse is. A wellestablished engineering principle is to build new structures to meet specific requirements by rearranging proven, existing components, as in mechanical structures and electronic circuits. The evolution of proteins by domain accretion and shuffling is one example of an analogous biological process. Mixing functional domains in new combinations is far more likely to produce a protein with novel activities than is the modification of one amino acid at a time. Single amino acid changes are more suitable for modulating existing functional properties (for example, ligand binding and allosteric responses) than for generating capabilities that did not previously exist. In addition to the combinatorial search via shuffling of existing exons, further variability results from the formation of novel exons. We do not yet know a great deal about any biases that may exist in the exonisation process. If it is correct to postulate that new functional exons arise by the exaptation of segments of mobile DNA, such as SINE elements, then it will be worthwhile to investigate the coding content of these elements to see if there is any tendency favouring sequences that encode useful folded polypeptide structures. The second major aspect of evolutionary change by natural genetic engineering is that it generally takes place after an activating event which produces what McClintock called a 'genome shock'. Activating events include loss of food, infection and interspecific hybridisation - just the events that we can infer from the geological and genomic records have happened repeatedly. Episodic activation of natural genetic engineering functions means that alterations to the genome occur in bursts rather than as independent events.

Thus, novel adaptations that require changes at multiple locations in the genome can arise within a single generation and can produce progeny expressing all the changes at once. There is no requirement, as in conventional theory, that each individual change be beneficial by itself. The episodic occurrence of natural genetic engineering bursts also makes it very easy to understand the punctuated pattern of the geological record. Moreover, the nature of activating challenges provides a comprehensible link to periodic disruptions in earth history. Geological upheavals that perturb an existing ecology are likely to lead to starvation, alteration of hostparasite relationships and unusual mating events between individuals from depleted populations.

A particular instance of the potential for stress-activated natural genetic engineering to produce complex novelties is the exaptation of an existing functional network following its duplication by WGD. Domains may be added to various proteins in the network to allow them to interact with a novel set of input and output molecules. In addition, insertions of connected regulatory signals at the cognate coding regions can generate a new transcriptional control circuit that may allow the modified network to operate under different conditions from its progenitor. The idea that genomic restructuring events may be integrated functionally in order to operate coordinately at a number of distinct loci encoding components of a regulatory network may seem extremely unlikely. However, the basic requirement for such integration is the ability to target DNA changes to co-regulated regions of the genome. Precisely this kind of targeting has been demonstrated for mobile elements in yeast, where retrotransposon integration activities interact with transcription or chromatin factors, and in Drosophila, where P elements can be engineered to home in on loci regulated by particular regulatory proteins. In addition, we know that mobile element insertion can be coupled with replication and DNA restructuring with transcription. Of course, the feasibility of such multi-locus functional integration of genome changes remains to be demonstrated in the laboratory. Fortunately, the experiments are straightforward; we can use appropriately engineered transposons and retrotransposons to search for coordinated multilocus mutations after activation. Clearly, the subject of functionally targeted changes to the genome belongs on the 21st century mobile DNA research agenda.

24.4.6   Future perspectives

Obligate intracellular bacteria are most often viewed as excellent model systems to understand the stable, symbiotic interactions that occur between prokaryotes and their eukaryotic hosts. The genomes of two strains of the aphid symbiont *Buchnera aphidicola* that are estimated to have diverged 50 to 70 million years ago show no indication of horizontal gene transfers, and many other obligate species have lost genes that encode DNA-repair and recombinase functions. Therefore, just as the presence of horizontal gene transfer is a hallmark of genome evolution in bacteria with free-living lifestyles, its absence has become a hallmark of genome evolution in bacteria with obligate intracellular lifestyles.

Are these specialised bacteria sealed to a fate of perpetual gene loss with no gene inflow by horizontal gene transfer? The answer depends, in part, on the lifestyle of the obligate species. Rates of horizontal gene transfer are affected by the rate of exposure to novel gene pools and, therefore, variations in transmission routes and host range will have a direct effect on rates of contact with foreign DNA. In support of this hypothesis, the obligate intra cellular species with mobile DNA are those that host-switch (for example, *Wolbachia*, *Coxiella* and *Phytoplasma* species). Even the Chlamydiaceae family, which has low genomic mobile-DNA contents, has extra chromosomal bacteriophages that infect the *Chlamydophila* genus. These data indicate that mobile genetic elements and horizontal gene transfer will not always be revealed by whole genome sequences, but might require the direct isolation of extrachromosomal elements.

The presence of mobile DNA alone does not specify whether a lateral transfer event has occurred in the recent or ancient evolutionary past. Instead, comparative sequence analyses of the mobile-DNA genes are the biologist's tool kit to reconstruct these evolutionary events. In many cases, the mobile DNA genes are disrupted by truncations and stop codons that reflect the ongoing degenerative processes in the genomes of intracellular bacteria.

However, in other cases, it is clear that recent horizontal transfer of mobile DNA has occurred in obligate intracellular bacteria, particularly in cases associated with bacteriophages. The recent discovery of mobile-DNA genes and horizontally acquired mobile DNA in host-switching obligate intracellular species argues for a new set of systemsbiology methods that should complement the studies of single bacteria but also take into account the genetic interplay of multiple micro-organisms that co-infect the same intracellular niche. Horizontal gene transfer could constitute both a serious threat to the stability of a highly integrated eukaryotic–prokaryotic association or a central source of evolutionary innovation to bacterial genomes that are otherwise experiencing extensive genome degeneration.

24.4.7   21st century view of evolutionary change

The evolutionary process has greatly expanded, thanks to studies of mobile DNA. Laboratory studies of plasmids, transposons, retrotransposons, NHEJ systems, reverse transcription, antigenic variation in prokaryotic and eukaryotic pathogens, lymphocyte rearrangements and genome reorganisation in ciliated protozoa have all made it possible to provide mechanistic explanations for events documented in the historical DNA record. We know that processes similar to those we document in our experiments have been major contributors to genome change in evolution. Using our knowledge of genome restructuring mechanisms, we can generate precise models to account for many duplications, amplifications, dispersals and rearrangements observed at both the genomic and proteomic levels. The genome DNA record also bears witness to sudden changes that affect multiple characters at once: horizontal transfer of large DNA segments, cell fusions and WGDs. These data are not readily compatible with earlier gradualist views on the nature of evolutionary variation. However, we are now able to apply the results of findings on the regulation of natural genetic engineering functions in the laboratory and in the field to make sense of the DNA record. Cell fusions and WGDs are events we know to activate DNA restructuring functions. Thus, it is not surprising that bursts of intracellular horizontal transfer, genome reduction and genome rearrangement follow these initial abrupt changes in the cells DNA. How a newly symbiotic cell or one with a newly doubled genome manages the transition to a stable genome structure that replicates and transfers reliably at cell division is another important subject for future research. The lessons we learn about silencing mobile DNA by internal deletion and RNA-directed chromatin modification are likely to prove helpful starting points.

Although there remain many gaps in our knowledge, we are now in a position to outline a distinctively 21st century scenario for evolutionary change. The scenario includes the following elements:

1. Hereditary variation arises from the non-random action of built-in biochemical systems that mobilise DNA and carry out natural genetic engineering.

2. Major disruptions of an organisms ecology trigger cell and genome restructuring. The ecological disruptions can act directly, through stress on individuals, or indirectly, through changes in the biota that favour unusual interactions between individuals (cell fusions, interspecific hybridisations). Triggering events continue until a new ecology has emerged that is filled with organisms capable of utilising the available resources.

3. Ecologically-triggered cell and genome restructurings produce organisms which, at some frequency, will possess novel adaptive features that suit the altered environment. Novel adaptive features can be complex from the beginning because they result from processes that operate on pre-existing functional systems, whose components can be amplified and rearranged in new combinations. Competition for resources (purifying selection) serves to eliminate those novel system architectures that are not functional in the new ecology.

4. Once ecological stability has been achieved, natural genetic engineering functions are silenced, the tempo of innovation abates, and microevolution can occur to fine-tune recent evolutionary inventions through successions of minor changes.

This 21st century view of evolution establishes a reasonable connection between ecological changes, cell and organism responses, widespread genome restructuring, and the rapid emergence of adaptive inventions. It also answers the objections to conventional theory raised by intelligent design advocates, because evolution by natural genetic engineering has the capacity to generate complex novelties. In other words, our best defense against anti-science obscurantism comes from the study of mobile DNA because that is the subject that has most significantly transformed evolution from natural history into a vibrant empirical science.

Molecular phylogenetics

23.1   Introduction

Molecular phylogenetics is the branch of phylogeny that analyses hereditary molecular differences, mainly in DNA sequences, to gain information on an organisms evolutionary relationships. The result of a molecular phylogenetic analysis is expressed in a phylogenetic tree. Molecular phylogenetics is one aspect of molecular systematics, a broader term that also includes the use of molecular data in taxonomy and biogeography.

23.2   Techniques and applications

Every living organism contains DNA, RNA, and proteins. In general, closely related organisms have a high degree of agreement in the molecular structure of these substances, while the molecules of organisms distantly related usually show a pattern of dissimilarity. Conserved sequences, such as mitochondrial DNA, are expected to accumulate mutations over time, and assuming a constant rate of mutation, provides a molecular clock for dating divergence. Molecular phylogeny uses such data to build a 'relationship tree' that shows the probable evolution of various organisms. Not until recent decades, however, has it been possible to isolate and identify these molecular structures.

The most common approach is the comparison of homologous sequences for genes using sequence alignment techniques to identify similarity. Another application of molecular phylogeny is in DNA barcoding, wherein the species of an individual organism is identified using small sections of mitochondrial DNA or chloroplast DNA.

Another application of the techniques that make this possible can be seen in the very limited field of human genetics, such as the ever-more-popular use of genetic testing to determine a childs paternity, as well as the emergence of a new branch of criminal forensics focused on evidence known as genetic fingerprinting.

A comprehensive step-by-step protocol on constructing phylogenetic tree, including DNA/Amino Acid contiguous sequence assembly, multiple sequence alignment, model-test (testing best-fitting substitution models) and phylogeny reconstruction using Maximum Likelihood and Bayesian Inference, is available at Nature Protocol.

23.3 Theoretical background

Early attempts at molecular systematics were also termed as chemotaxonomy and made use of proteins, enzymes, carbohydrates, and other molecules that were separated and characterised using techniques such as chromatography. These have been replaced in recent times largely by DNA sequencing, which produces the exact sequences of nucleotides or bases in either DNA or RNA segments extracted using different techniques. In general, these are considered superior for evolutionary studies, since the actions of evolution are ultimately reflected in the genetic sequences. At present, it is still a long and expensive process to sequence the entire DNA of an organism (its genome), and this has been done for only a few species. However, it is quite feasible to determine the sequence of a defined area of a particular chromosome. Typical molecular systematic analyses require the sequencing of around 1000 base pairs. At any location within such a sequence, the bases found in a given position may vary between organisms. The particular sequence found in a given organism is referred to as its haplotype. In principle, since there are four base types, with 1000 base pairs, we could have 41000 distinct haplotypes. However, for organisms within a particular species or in a group of related species, it has been found empirically that only a minority of sites show any variation at all and most of the variations that are found are correlated, so that the number of distinct haplotypes that are found is relatively small.

In a mack molecular systematic analysis, the haplotypes are determined for a defined area of genetic material; a substantial sample of individuals of the target species or other taxon is used, however many current studies are based on single individuals. Haplotypes of individuals of closely related, but different, taxa are also determined. Finally, haplotypes from a smaller number of individuals from a definitely different taxon are determined. These are referred to as an out group. The base sequences for the haplotypes are then compared. In the simplest case, the difference between two haplotypes is assessed by counting the number of locations where they have different bases. This is referred to as the number of substitutions (other kinds of differences between haplotypes can also occur, for example the insertion of a section of nucleic acid in one haplotype that is not present in another). The difference between organisms is usually re-expressed as a percentage divergence, by dividing the number of substitutions by the number of base pairs analysed: the hope is that this measure will be independent of the location and length of the section of DNA that is sequenced.

An older and superseded approach was to determine the divergences between the genotypes of individuals by DNA-DNA hybridisation. The advantage claimed for using hybridisation rather than gene sequencing was that it was based on the entire genotype, rather than on particular sections of DNA.

Modern sequence comparison techniques overcome this objection by the use of multiple sequences.

23.4   Mechanism of change

The mechanisms of change include:

1. Random mutation, which can be seen as genetic drift in the absence of selective pressure.
2. Sequence duplication, which may be duplication of small segments, genes or even whole genomes.
3. Recombination, which includes transposons, translocations and viral activity, to mix up sequences within an organism, to remove sequences or to introduce sequences from another organism.

There is considerable debate regarding the best approach to take when analysing sequence alignments for phylogenetic relationships. Accepted approaches include distance calculations, parsimony and maximum likelihood. New on the scene is Bayesian analysis, which is expected to gain popularity as people become familiar with it. It is good to become familiar with the different methods of analysis. It helps in understanding the arguments being made, both in terms of how things should be done and of the resulting analyses. When examining the results of molecular phylogenetic analysis, care should be taken to compare the results to other independent means of analysis and/or to other data sets.

23.4.1   Associated concepts

The associated concepts in phylogenetic analysis are:

1. Phylogenetics vs taxonomy.
2. Cladistic vs phenetic.
3. Clustering.
4. Parsimony vs maximum likelihood.

Charles Darwin was the first to recognise that the systematic hierarchy represented a rough approximation of evolutionary history. However, it was not until the 1950s that the German entomologist Willi Hennig proposed that systematics should reflect the known evolutionary history of lineages as closely as possible, an approach he called phylogenetic systematics.

The followers of Hennig were referred to as 'cladists' by his opponents, because of the emphasis on recognising, only monophyletic groups, a group plus all of its descendants or clades. However, the cladists quickly adopted that term as a helpful label and nowadays, cladistic approaches to systematics are used routinely.

Phylogenetic systematics

Phylogenetic systematics is that field of biology that does deal with identifying and understanding the evolutionary relationships among the many different kinds of life on earth, both living (extant) and dead (extinct). Evolutionary theory states that similarity among individuals or species is attributable to common descent, or inheritance from a common ancestor. Thus, the relationships established by phylogenetic systematics often describe a species evolutionary history and hence, it is phylogeny, the historical relationships among lineages or organisms or their parts, such as their genes.

23.5 Evolutionary process

Following points will facilitate the discussion of evolutionary process.

23.5.1 Genetic variation: changes in a gene pool

Evolution is not always discrete with clearly defined boundaries that pinpoint the origin of a new species, nor is it a steady continuum. Evolution requires genetic variation, which results from changes within a gene pool, the genetic make-up of a specific population. A gene pool is the combination of all the alleles; alternative forms of a genetic locus; for all traits that population may exhibit. Changes in a gene pool can result from mutation-variation within a particular gene; or from changes in gene frequency; the proportion of an allele in a given population.

23.5.2 Occurrence of genetic variation

Every organism possesses a genome that contains all of the biological information needed to construct and maintain a living example of that organism. The biological information contained in a genome is encoded in the nucleotide sequence of its DNA or RNA molecules and is divided into discrete units called genes. The information stored in a gene is read by proteins, which attach to the genome and initiate a series of reactions called gene expression.

Every time a cell divides, it must make a complete copy of its genome, a process called DNA replication. DNA replication must be extremely accurate to avoid introducing mutations or changes in the nucleotide sequence of a short region of the genome. Inevitably, some mutations do occur, usually in one of two ways; either from errors in DNA replication or from damaging effects of chemical agents or radiation that react with DNA and change the structure of individual nucleotides. Many of these mutations result in a change that has no effect on the functioning of the genome, referred to as silent mutations. Silent mutations include virtually all changes that happen in the noncoding components of genes and gene-related sequences.

Mutations in the coding regions of genes are much more important. Here we must consider the importance of the same mutation in a somatic cell compared with a germ line cell. A somatic cell is any cell of an organism other than a reproductive cell, such as a sperm or egg cell. (A germ cell line is any line of cells that gives rise to gametes and is continuous through the generations.) Because a somatic cell does not pass on copies of its genome to the next generation, a somatic cell mutation is important only for the organism in which it occurs and has no potential evolutionary impact. In fact, most somatic mutations have no significant effect because there are many other identical cells in the same tissue.

On the other hand, mutations in germ cells can be transmitted to the next generation and will then be present in all of the cells of an individual who inherits that mutation. Even still, mutations within germ line cells may not change the phenotype of the organism in any significant way. Those mutations that do have an evolutionary effect can be divided into two categories, loss-of-function mutations and gain-of-function mutations. A loss-of-function mutation results in reduced or abolished protein function. Gain-of-function mutations, which are much less common, confer an abnormal activity on a protein.

The randomness with which mutations can occur is an important concept in biology and is a requirement of the Darwinian view of evolution, which holds that changes in the characteristics of an organism occur by chance and are not influenced by the environment in which the organism lives. Beneficial changes within an organism are then positively selected for, whereas harmful changes are negatively selected.

23.5.3   Drivers of evolution: selection, drift and founder effects

The new alleles appear in a population because of mutations that occur in the reproductive cells of an organism. This means that many genes are polymorphic, that is, two or more alleles for that gene are present in a population. Each of these alleles has its own allele or gene frequency, a measure of how common an allele is in a population. Allele frequencies vary over time because of two conditions, natural selection and random drift.

Natural selection

Natural selection is the process whereby one genotype, the hereditary constitution of an individual, leaves more offspring than another genotype because of superior life attributes, termed fitness. Natural selection acts on genetic variation by conferring a survival advantage to those individuals harbouring a particular mutation that tends to favour a changing environmental condition. These individuals then reproduce and pass on this 'new' gene, altering their gene pool. Natural selection, therefore, decreases the frequencies of alleles that reduce the fitness of an organism and increases the frequency of alleles that improve fitness. Thus, 'natural selection' is the principle by which each slight variation, if useful, is preserved. It is important to point out that natural selection does not always represent progress, only adaptation to a changing surrounding, that is evolution attributable to natural selection is devoid of intent something does not evolve to better itself, only to adapt. Because environments are always changing, what was once an advantageous mutation can often become a liability further down the evolutionary line.

Random drift

The term random drift actually encompasses a number of distinct processes, sometimes referred to as outcomes. They include indiscriminate parent sampling, the founder effect and fluctuations in the rate of evolutionary processes such as selection, migration and mutation. Parent sampling is the process of determining which organisms of one generation will be the parents of the next generation. Parent sampling may be discriminate, that is, with regard to fitness differences or indiscriminate, without regard to fitness differences. Discriminate parent sampling is generally considered natural selection, whereas indiscriminate parent sampling is considered random drift.

23.6   Sampling

Suppose a population of red and brown squirrels share a habitat with a colour blind predator. Although the predator is colour blind, the brown squirrels seem to die in greater numbers than the red squirrels, suggesting that the brown squirrels just seem to be unlucky enough to come into contact with the predator more often. As a result, the frequency of brown squirrels in the next generation is reduced. More red squirrels survive to reproduce or are sampled, but it is without regard to any differences in fitness between the two groups. The physical differences of the groups do not play a causal role in the differences in reproductive success. Now, let's say that the predator is not colour blind and can now see the red squirrels better than the brown squirrels, resulting in a better survival rate for the brown squirrels. This would be a case of discriminate parent sampling or natural selection.

23.7   Founder effect

Another important cause of genetic drift is the founder effect, the difference between the gene pool of a population as a whole and that of a newly isolated population of the same species. The founder effect occurs when population are started from a small number of pioneer individuals of one original population. Because of small sample size, the new population could have a much different genetic ratio than the original population. An example of the founder effect would be when a plant population results from a single seed. Thus far, we have discussed natural selection and random drift as events that occur in isolation from one another. However, in most population, the two processes will be occurring at the same time. Furthermore, there is great debate over whether, in particular instances and in general, natural selection is more prevalent than random drift.

23.8   Phylogenetic trees: Presenting evolutionary relationships

Systematics describes the pattern of relationships among taxa and is intended to help us understand the history of all life. But history is not something we can see it has happened once and leaves only clues as to the actual events. Scientists use these clues to build hypotheses, or models, of lifes history. In phylogenetic studies, the most convenient way of visually presenting evolutionary relationships among a group of organisms is through illustrations called phylogenetic trees.

This can be either an existing species or an ancestor:

1. Branch: defines the relationship between the taxa in terms of descent and ancestry.
2. Branch length: represents the number of changes that have occurred in the branch.
3. Topology: the branching patterns of the tree.
4. Root: the common ancestor of all taxa.
5. Distance scale: scale that represents the number of differences between organisms or sequences.
6. Clade: a group of two or more taxa or DNA sequences that includes both their common ancestor and all of their descendants.
7. Operational taxonomic unit (OTU): taxonomic level of sampling selected by the user to be used in a study, such as individuals, populations, species, genera or bacterial strains.

A phylogenetic tree is composed of nodes, each representing a taxonomic unit (species, populations, individuals) and branches, which define the relationship between the taxonomic units in terms of descent and ancestry. Only one branch can connect any two adjacent nodes. The branching pattern of the tree is called the topology, and the branch length usually represents the number of changes that have occurred in the branch. This is called a scaled branch. Scaled trees are often calibrated to represent the passage of time. Such trees have a theoretical basis in the particular gene or genes under analysis.

Branches can also be unscaled, which means that the branch length is not proportional to the number of changes that has occurred, although the actual number may be indicated numerically somewhere on the branch. Phylogenetic trees may also be either rooted or unrooted. In rooted trees, there is a particular node, called the root, representing a common ancestor, from which a unique path leads to any other node. An unrooted tree only specifies the relationship among species, without identifying a common ancestor, or evolutionary path. Phylogenetic trees, a convenient way of representing evolutionary relationships among a group of organisms can be drawn in various ways. Branches on phylogenetic trees may be scaled representing the amount of evolutionary change, time or both, when there is a molecular clock or they may be unscaled and have no direct correspondence with either time or amount of evolutionary change. Phylogenetic trees may be rooted or unrooted. In the case of unrooted trees, branching relationships between taxa are specified by the way they are connected to each other, but the position of the common ancestor is not. For example, on an unrooted tree with five species, there are five branches (four external, one internal) on which the tree can be rooted. Rooting on each of the five branches has different implications for evolutionary relationships.

23.9 Methods of phylogenetic analysis

Two major groups of analyses exist to examine phylogenetic relationships: phenetic methods and cladistic methods. It is important to note that phenetics and cladistics have had an uneasy relationship over the last 40 years or so. Most of today's evolutionary biologists favour cladistics, although a strictly cladistic approach may result in counterintuitive results.

23.9.1 Phenetic method of analysis

Phenetics, also known as numerical taxonomy, involves the use of various measures of overall similarity for the ranking of species. There is no restriction on the number or type of characters (data) that can be used, although all data must be first converted to a numerical value, without any character 'weighting'. Each organism is then compared with every other for all characters measured and the number of similarities (or differences) is calculated.

The organisms are then clustered in such a way that the most similar are grouped close together and the more different ones are linked more distantly. The taxonomic clusters, called phenograms, that result from such an analysis do not necessarily reflect genetic similarity or evolutionary relatedness. The lack of evolutionary significance in phenetics has meant that this system has had little impact on animal classification and as a consequence, interest in and use of phenetics has been declining in recent years.

23.9.2   Cladistic method of analysis

An alternative approach to diagramming relationships between taxa is called cladistics. The basic assumption behind cladistics is that members of a group share a common evolutionary history. Thus, they are more closely related to one another than they are to other groups of organisms. Related groups of organisms are recognised because they share a set of unique features (apomorphies) that were not present in distant ancestors but which are shared by most or all of the organisms within the group. These shared derived characteristics are called synapomorphies. Therefore, in contrast to phenetics, cladistics groupings do not depend on whether organisms share physical traits but depend on their evolutionary relationships. Indeed, in cladistic analysis two organisms may share numerous characteristics but still be considered members of different groups.

Cladistic analysis entails a number of assumptions. For example, species are assumed to arise primarily by bifurcation or separation, of the ancestral lineage; species are often considered to become extinct upon hybridisation (crossbreeding) and hybridisation is assumed to be rare or absent. In addition, cladistic groupings must possess the following characteristics: all species in a grouping must share a common ancestor and all species derived from a common ancestor must be included in the taxon. The application of these requirements results in the following terms being used to describe the different ways in which groupings can be made:

1. A monophyletic grouping is one in which all species share a common ancestor, and all species derived from that common ancestor are included. This is the only form of grouping accepted as valid by cladists.

2. A paraphyletic grouping is one in which all species share a common ancestor, but not all species derived from that common ancestor are included. A polyphyletic grouping is one in which species that do not share an immediate common ancestor are lumped together, while excluding other members that would link them.

23.10   Origins of molecular phylogenetics

Macromolecular data, meaning gene (DNA) and protein sequences, are accumulating at an increasing rate because of recent advances in molecular biology. For the evolutionary biologist, the rapid accumulation of sequence data from whole genomes has been a major advance, because the very nature of DNA allows it to be used as a 'document' of evolutionary history. Comparisons of the DNA sequences of various genes between different organisms can tell a scientist a lot about the relationships of organisms that cannot otherwise be inferred from morphology or an organisms outer form and inner structure.

Because genomes evolve by the gradual accumulation of mutations, the amount of nucleotide sequence difference between a pair of genomes from different organisms should indicate how recently those two genomes shared a common ancestor. Two genomes that diverged in the recent past should have fewer differences than two genomes whose common ancestor is more ancient. Therefore, by comparing different genomes with each other, it should be possible to derive evolutionary relationships between them, the major objective of molecular phylogenetics. Molecular phylogenetics attempts to determine the rates and patterns of change occurring in DNA and proteins and to reconstruct the evolutionary history of genes and organisms. Two general approaches may be taken to obtain this information. In the first approach, scientists use DNA to study the evolution of an organism. In the second approach, different organisms are used to study the evolution of DNA. Whatever the approach, the general goal is to infer process from pattern: the processes of organismal evolution deduced from patterns of DNA variation and processes of molecular evolution inferred from the patterns of variations in the DNA itself.

23.10.1 Molecular phylogenetic analysis

Fundamental elements as we just discussed, macromolecules, especially gene and protein sequences, have surpassed morphological and other organismal characters as the most popular forms of data for phylogenetic analysis. Therefore, this next section will concentrate only on molecular data. It is important to point out that a single, all-purpose recipe does not exist for phylogenetic analysis of molecular data. Although numerous algorithms, procedures, and computer programmes have been developed, their reliability and practicality are, in all cases, dependent upon the size and structure of the dataset under analysis. The merits and shortfalls of these various methods are subject to much scientific debate, because the danger of generating incorrect results is greater in computational molecular phylogenetics than many other fields of science. Occasionally, the limiting factor in such analyses is not so much the computational method used, but the users understanding in of what the method is actually doing with the data. Therefore, the goal of this section is to demonstrate to the reader that practical analysis should be thought of both as a search for a correct model (analysis) as well as a search for the correct tree (outcome). Phylogenetic tree-building models presume particular evolutionary models. For any given set of data, these models may be violated because of various occurrences, such as the transfer of genetic material between organisms. Therefore, when interpreting a given analysis, a person should always consider the model used and entertain possible explanations for the results obtained. For example, models used in molecular phylogenetic analysis.

Methods make 'default' assumptions

The sequence is correct and originates from the specified source:

1. The sequences are homologous — all descended in some way from a shared ancestral sequence.
2. Each position in a sequence alignment is homologous with every other in that alignment.
3. Each of the multiple sequences included in a common analysis has a common phylogenetic history with the other sequences.
4. The sampling of taxa is adequate to resolve the problem under study.
5. Sequence variation among the samples is representative of the broader group.
6. The sequence variability in the sample contains phylogenetic signal adequate to resolve the problem under study.

23.10.2 Four steps of phylogenetic analysis

A straightforward phylogenetic analysis consists of four steps:

1. Alignment-building the data model and extracting a dataset.
2. Determining the substitution model-consider sequence variation.
3. Tree building.
4. Tree evaluation.

Tree building

Key features of DNA-based phylogenetic trees: Studies of gene and protein evolution often involve the comparison of homologs, sequences that have common origins but may or may not have common activity. Sequences that share an arbitrary level of similarity determined by alignment of matching bases are homologous. These sequences are inherited from a common ancestor that possessed similar structure, although the ancestor may be difficult to determine because it has been modified through descent.

Homologs are most commonly defined as orthologs, paralogs or xenologs. Orthologs are homologs produced by speciation. They represent genes derived from a common ancestor that diverged because of divergence of the organism. Orthologs tend to have similar function.

Paralogs are homologs produced by gene duplication and represent genes derived from a common ancestral gene that duplicated within an organism and then diverged. Paralogs tend to have different functions. Xenologs are homologs resulting from the horizontal transfer of a gene between two organisms. The function of xenologs can be variable, depending on how significant the change in context was for the horizontally moving gene. In general, though, the function tends to be similar. A typical gene-based phylogenetic tree is shown in Fig. 23.1. This tree shows the relationship between four homologous genes: A, B, C and D.

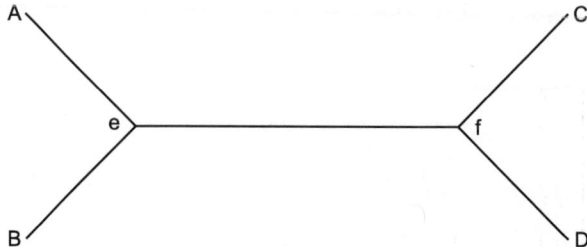

Figure 23.1: Gene based phylogenetic tree.

The topology of this tree consists of four external nodes (A, B, C and D), each one representing one of the four genes and two internal nodes (e and f) representing ancestral genes. The branch lengths indicate the degree of evolutionary differences between the genes. This particular tree is unrooted it is only an illustration of the relationships between genes A, B, C and D and does not signify anything about the series of evolutionary events that led to these genes. A rooted tree is often referred to as an inferred tree.

This is to emphasise that this type of illustration depicts only the series of evolutionary events that are inferred from the data under study and may not be the same as the true tree or the tree that depicts the actual series of evolutionary events that occurred.

To distinguish between the pathways, the phylogenetic analysis must include at least one outgroup, a gene that is less closely related to A, B, C and D than these genes are to each other (panel below). Outgroups enable the root of the tree to be located and the correct evolutionary pathway to be identified. Let's say that the four case, an outgroup could be a gene from another primate, such as baboon, which is known to have branched away from the four species above before the common ancestor of the species. Homologous genes used in the previous tree examples come from human, chimpansee, gorilla and orangutan. In this case, an outgroup could be a gene from another primate, such as baboon, which is known to have branched away from the four species above before the common ancestor of the species.

23.11 Gene trees versus species trees—Why are they different?

It is assumed that a gene tree, because it is based on molecular data, will be a more accurate and less ambiguous representation of the species tree that is obtainable by morphological comparisons. This may indeed be the case, but it does not mean that the gene tree is the same as the species tree. For this to be true, the internal nodes in both trees would have to be precisely equivalent, and they are not. An internal node in a gene tree indicates the divergence of an ancestral gene into two genes with different DNA sequences, usually resulting from a mutation of one sort or another.

An internal node in a species tree represents what is called a speciation event, whereby the population of the ancestral species splits into two groups that are no longer able to interbreed. These two events, mutation and speciation, do not always occur at the same time.

23.12   Molecular phylogenetics terminology

Monophyletic: Two or more DNA sequences that are derived from a single common ancestral DNA sequence.

Clade: A group of monophyletic DNA sequences that make up all of the sequences included in the analysis that are descended from a particular common ancestral sequence.

Parsimony: An approach that decides between different tree topologies by identifying the one that involves the shortest evolutionary pathway. This is the pathway that requires the smallest number of nucleotide changes to go from the ancestral sequence, at the root of the tree, to all of the present-day sequences that have been compared.

23.13   Molecular clock hypothesis

Molecular clock hypothesis states that nucleotide substitutions, or amino acid substitutions in proteins are being compared, occur at a constant rate, that is, the degree of difference between two sequences can be used to assign a date to the time at which their ancestral sequence diverged. The rate of molecular change differs among groups of organisms, among genes, and even among different parts of the same gene. Furthermore, molecular clocks require calibration with fossils to determine timing of origin of clades and thus their accuracy is crucially dependent on the fossil record, or lack thereof, for the groups under study. Fossil DNA older than about 25,000–50,000 years is virtually empty of phylogenetic signal except in rare instances and therefore traditional morphological studies of extinct and extant organisms remain a crucial component of phylogenetic analysis.

23.14   Importance of molecular phylogenetics

The field of molecular phylogenetics has grown, both in size and in importance, since its inception in the early 1990s, attributable mostly to advances in molecular biology and more rigorous methods for phylogenetic tree building. The importance of phylogenetics has also been greatly enhanced by the successful application of tree reconstruction, as well as other phylogenetic techniques. Today, a survey of the scientific literature will show that molecular biology, genetics, evolution, development, behaviour, epidemiology, ecology, systematics, conservation biology and forensics are but a few examples of perplexing issues in the many disparate fields conceptually united by the methods and theories of molecular phylogenetics. Phylogenies are used essentially the same way in all of these fields, either by drawing inferences from the structure of the tree or from the way the character states map onto the tree. Biologists can then use these clues to build hypotheses and models of important events in history. Broadly speaking, the relationships established by phylogenetic trees often describe a species' evolutionary history and hence, its phylogeny; the historical relationships among lineages or organisms or their parts, such as their genes. Phylogenies may be thought of as a natural and meaningful way to order data, with an enormous amount of evolutionary information contained within their branches. Scientists working in these different areas can then use these phylogenies to study and elucidate the biological processes occurring at many levels of life's hierarchy.

23.15 Common tree-building methods used in phylogenetic inference

Phylogenetic inference can be defined as the process of determining the estimated evolutionary history by analysis of a given data set. Increasingly, molecular data sets, such as DNA and protein sequences, are used to develop these phylogenies. The reasons for building a phylogenetic tree are as diverse as the methods used to produce the trees. The process of phylogenetic analysis can be summarised by the following steps.

The first two steps are preparatory for the subsequent steps that involve tree building and evaluation of the resultant tree.

1. The first step is the alignment of either the nucleotide or the amino acid sequences for the taxa of interest. It is generally agreed upon that amino acid sequences produce a tree closest to the true tree. This is due to the higher rate of conservation of amino acid sequences and protein structure. Manual alignment editing is recommended over fully computational multiple alignments as the algorithms and programmes are not yet optimal for phylogenetic alignment. Regardless of the method for alignment, the final alignment should be carefully scrutinised with any independent phylogenetic evidence and other assumptions of structure and function. Once one proceeds to tree building, the computer-generated alignment will be blind to any errors in alignment.

2. The second step will be to determine the presence of a phylogenetic signal. Most of the sequence analyses fall between two extremes: identical sequences and sequences which have become so divergent as to become randomised in relation to the phylogenetic history. The former case dictates no further analysis. The latter will result in an inferred phylogeny, though the randomness of the resulting phylogeny may not be worth the effort. Those sequence alignments that fall in between will have a mixture of conserved and random positions and will be the most useful in phylogenetic inference.

Once the alignment is complete, the next steps in phylogenetic inference are to decide the most appropriate tree-building method for a specified data set followed by choosing a strategy to find the best tree under the selected optimality criterion. Finally, the tree obtained must be scrutinised to determine the level of confidence that can be placed on the results. One of the most complex issues faced during the process of phylogenetic inference is choosing the tree-building method.

23.15.1 Classification of tree-building methods

Tree-building methods can be classified in two ways.

The first way to classify these methods is to define them as either algorithm-based or criterion-based

Though the procedure involved in each of these methods is different, the same algorithm could potentially be used in either method. An algorithm-based method generates a tree by following a series of steps, whereas criterion-based methods define an optimality criterion for comparing alternative phylogenies to one another and deciding, which one is better. Therefore, there is a big advantage when working with criteria-based methods because scores are assigned to every examined tree and can be used to rank the resultant phylogenies in order of preference. This provides, the user with immediate knowledge about the strength of support for that tree. Strictly algorithmic methods are computationally much faster than the criteria-based methods, because they do not require evaluation of a large number of competing trees. Due to the large number of possible solutions, criteria-based methods do not produce exact results for data sets with more than 8–20 taxa.

Alternatively, tree-building methods can be classified into distance-based versus character-based methods

A distance-based method computes pairwise distances according to some measure. Then, the actual data is discarded and the fixed distances are used in the derivation of trees. Trees derived by way of a character-based method have been optimised according to the distribution of actual data patterns in relation to a specified character.

Cluster analysis and neighbour joining are examples of methods defined solely on the basis of an algorithm or of methods that are unable to separate the task of finding an optimal tree from that of evaluating a specific tree, unlike the criteria-based methods. Cluster analysis constructs trees by linking the least distant pairs of taxa, followed by successively more distant taxa or groups of taxa. Once two taxa are linked, they lose their individual identities and are subsequently referred to as a single cluster. Neighbour joining is related to this traditional cluster analysis except it removes the assumption that data are ultrameric. The tree is constructed by linking the least distant pairs of nodes as defined by a modified matrix. The modified distance matrix is constructed by adjusting the separation between each pair of nodes on the basis of the average divergence from all other nodes.

There are three common types of optimality criteria that will be briefly discussed including parsimony, likelihood and pairwise distance. Both maximum parsimony and maximum likelihood use the original data set for inference. Maximum parsimony falls under the philosophy of 'the simpler hypotheses are preferable to the complicated ones'. It works in such a way as to choose from the alternative trees, the one with the fewest character-state transformations, thus minimising homoplasy (e.g. convergence, reversal). Thus, this optimality criterion operates by selecting trees that minimise the total tree length. This method tends to yield numerous trees with the same score, which is not characteristic of other methods such as distance or maximum likelihood. Parsimony is less dependent upon assumptions about the sequence evolution than some other methods and amenable to weighting in order to accommodate any substitution bias. The drawbacks for parsimony include slowed computation with weighting and poor performance when there is substantial among-site rate heterogeneity. Though there are several modifications that can correct for heterogeneity, such as modifying the data set or reweighting positions according to their propensity to change (successive approximation), this could potentially lead to errors in a prior step if the preliminary tree contains any errors.

An area of trouble using maximum parsimony is referred to as the Felsenstein zone. This zone is created when there exists strongly unequal rates of change along different branches of a tree or even with equal rates of change in cases of long-branch attraction. Long-branch refers to a lineage that evolved so much between nodes in the phylogeny that its character states have been effectively randomised with respect to the other taxa. Once in the Felsenstein zone misleading inferences are produced. At the ends of these long branches, character states are exhibited that no longer retain genealogical information leading to a distortion in the inference. Particularly deceptive are taxa on long branches that have converged on character states present in other taxa within the analysis. This appears as a false phylogenetic signal, obscuring the true signal. There are several ways to fix this problem including weighted parsimony, the use of relative apparent synapomorphy analysis (RASA) or using maximum likelihood that incorporates models of evolutionary change.

Parsimony once took the lead as the most favoured method; however, maximum likelihood appears to be replacing parsimony, particularly as this method becomes better defined. The critical difference between these two methods is that parsimony minimises the amount of evolutionary change required for data explanation, while maximum likelihood attempts to estimate the actual amount of change according to the evolutionary model in place. Maximum likelihood works with a prior nucleotide substitution model to compute a likelihood score for each tree given the original data. Before beginning, either an evolutionary model must be specified that can account for the conversion of one sequence into another or parametres must be selected that can be estimated from the data. Then the maximum likelihood approach evaluates the probability that the selected evolutionary model will have generated the observed sequences. The trees yielding the highest likelihoods are used to infer phylogeny. The substitution model should be optimised to fit the observed data as modifying the substitution parametres modify the likelihood of the data associated with particular trees. The greatest drawback to using maximum likelihood is the vast amount of computation time required.

Maximum likelihood is not always available for use. An alternative method that also minimises the impact of the underestimation problem present in parsimony is the pairwise distance method. It works on the idea that corrected distances to account for superimposed changes can be obtained by estimating the number of unseen events using the same models used with maximum likelihood. The corrected distances are estimates of the true evolutionary distance. The drawback for this method is the loss of data during the process.

Because many of the more complex models require an enormous amount of time to complete the computations, heuristic methods are often selected in place of the alternative search methods. A heuristic method does not guarantee finding the optimal solution, but does provide a large increase in speed. Such is the case with maximum likelihood. Likelihood has several advantages including consistency, lower variance in estimation, and its robustness to violations of its assumptions, so many attempts have been made to incorporate maximum likelihood into a heuristic method that will simultaneously optimise the substitution model and the tree for a given data set.

In applying any of these methods, there are two key ideas that have been emphasised throughout the development of phylogenetics. The importance of the starting data cannot be stressed enough. The type of data not only determines which method to choose for analysis, but also determines the validity of the results obtained from a selected method. It follows, then, that the validity of the resulting tree is dependent upon the appropriateness of the model used in tree generation. Phylogenetic inference methods are under continual evaluation and improvement upon the accuracy and speed of computation in these methods is a continual process. In several cases, it has been determined that trees obtained from the simpler methods produce as good results as those obtained by more sophisticated methods. Though this may be an accurate assessment, the more complex methods have the advantage of producing several alternatives, which allow for the different topologies to be evaluated for statistical significance. There is no agreement, yet, on which method is the best method and more than likely there never will be as the best method is dependent upon the type of data with which one begins. As long as the method has a theoretical justification, then it is more important to choose a good gene or a large number of amino acids than it is to choose a particular tree-building method.

23.16   Phylogenetic prediction

23.16.1 Tree of life

Evolutionary trees

An evolutionary tree is a two-dimensional graph showing the evolutionary relationship among a set of items being compared. This set can be organisms, genes or DNA sequences. Consider for the moment that each of the units in the set are referred to as a taxon. Each taxon will be defined by a distinct unit on the tree. An evolutionary tree is composed of outer branches or leaves that represent the taxa and nodes and branches representing the relationships among the taxa. Two taxa that are derived from the same common ancestor will share a node in the graph. In general, approaches to designing evolutionary trees attempt are made to define the length of each branch to the next node according to the number of sequence level changes that occurred. One thing to be careful of in phylogenetic analysis is that this distance may not be in direct relation to evolutionary time. Analyses that prescribe to the theory of a uniform rate of mutation are known as the molecular clock hypothesis.

Rooted trees

In a rooted tree topology, one sequence (the root) is defined to be the common ancestor of all of the other sequences. A unique path leads from the root node to any other node and the direction of the path indicates evolutionary time. The root is chosen by including a sequence from an organism that is thought to have branched off earlier than the other sequences. If the molecular clock hypothesis holds, it is also possible to predict a root. As the number of sequences increase, the number of possible rooted trees increases very rapidly. In some cases, a bifurcating binary tree is the best model to simulate evolutionary events in which case one species branches off into two separate species. Example of a rooted tree is shown in Fig. 23.2.

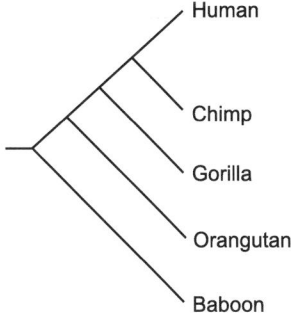

Figure 23.2: Example of a rooted tree.

Star topology (unrooted trees)

An unrooted tree (sometimes referred to as a star topology) shows the evolutionary relationship among sequences, without revealing the location of the oldest ancestry. There are fewer choices for an unrooted tree than a rooted tree. Example of an unrooted tree is shown in Fig. 23.3.

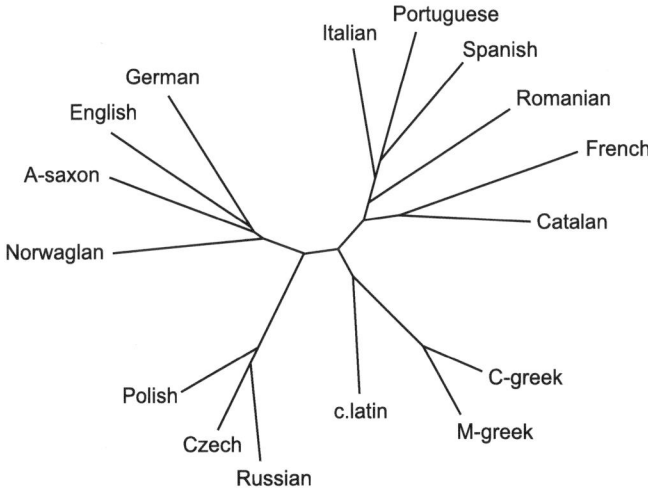

Figure 23.3: Example of an unrooted tree.

23.16.2 Methods for determining evolutionary trees

There are three methods used to calculate the tree(s) that best account for the observed variation in a set of sequences. These methods are maximum parsimony, distance and maximum likelihood.

Maximum parsimony

Maximum parsimony methods predict the evolutionary tree that minimises the number of steps required to generate the observed variation in the sequences. In order to construct a tree using maximum parsimony, a multiple sequence alignment must first be obtained. For each aligned position, phylogenetic trees that require the smallest number of evolutionary changes to produce the observed sequence changes are identified. This continues for each position in the alignment. Those trees that produce the smallest number of changes overall for all sequence positions are identified. This is a rather time consuming algorithm that only works well if the sequences have a strong sequence similarity. Consider the example above. There are a total of four sequences, (Fig. 23.4), which gives a possibility of three different unrooted trees as shown in Fig. 23.5. In this case some sites are informative and other sites are not. An informative site has the same sequence character in atleast two different sequences. Only the informative sites need to be considered.

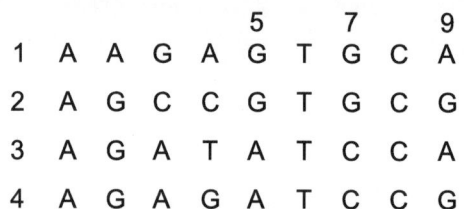

					5		7		9
1	A	A	G	A	G	T	G	C	A
2	A	G	C	C	G	T	G	C	G
3	A	G	A	T	A	T	C	C	A
4	A	G	A	G	A	T	C	C	G

Figure 23.4: Multiple alignment for phylogeny.

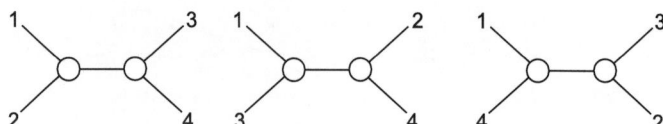

Figure 23.5: Possible trees from four sequences.

In this case, adding the number of changes at each informative site for each tree and picking the tree requiring the least total number of changes obtain the optimal tree. For a large number of sequences the number of trees to examine becomes so large that it might not be possible to examine all possible trees. Some programmes, such as PAUP, add features that will allow the user to invoke a heuristic that will keep representative trees that best fit the data.

The informative sites in the example of alignment are position 5, 7 and 9 in Fig. 23.4. Let's go through the possible trees and figure out the number of rearrangements for each in the informative sites. One problem with determining evolutionary distance between sequences is that columns representing greater variation dominate the analysis. In order to overcome this problem of determining long branch lengths is to look only at transversion events, which are the most significant base changes (i.e., changes a purine to a pyrimidine or *vice versa*). This is referred to as Lakes method of invariants.

Distance methods

The distance method for construction of phylogenetic trees looks at the number of changes between each pair in a group of sequences to produce a phylogenetic tree of the group. The goal of distance methods is to identify a tree that positions neighbours correctly and that also has branch lengths, which reproduce the original data as closely as possible. For phylogenetic analysis, the distance score counted as either the number of mismatched positions in the alignment or the number of sequence positions that must be changed to generate the other sequence is used. The success of distance methods depends on the degree to which the distances among a set of sequences can be made additive on a predicted evolutionary tree. The alignment is shown in Fig. 23.6.

A   ACGCGTTGGGCGATGGCAAC

B   ACGCGTTGGGCGACGGTAAT

C   ACGCATTGAATGATGATAAT

D   ACACATTGAGTGATAATAAT

Figure 23.6: The alignment.

The distances between these sequences are shown in Table 23.1.

Table 23.1: Showing distances between the above sequences.

	A	B	C	D
A	–	3	7	8
B	–	–	6	7
C	–	–	–	3
D	–	–	–	–

Note: Distances are nothing but the number of changes in bases/aminoacids between any two sequences under comparison.

Using this information, an unrooted tree showing the relationship between these sequences can be drawn as shown in Fig. 23.7.

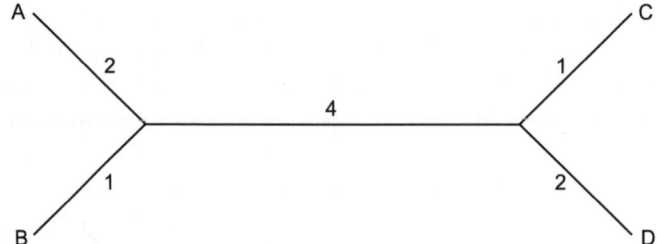

Figure 23.7: Unrooted tree resulting from the above four sequences.

Fitch and margoliash method

The Fitch and Margoliash method uses a distance table. The sequences are combined in trees to define the branches of the predicted tree and to calculate the branch lengths of the tree. The Fitch and Margoliash method can be extended to three or more sequences. The Fitch-Margoliash algorithm can be extended to three or more sequences.

Repeat the process until all lengths have been identified, in which case there is only single composite node left. Thus the steps involved in Fitch-Margoliash algorithm are summarised here pointwise.

1. Find the most closely related pairs of sequences (A, B).
2. Treat the rest of the sequences as a composite. Calculate the average distance from A to all others; and from B to all others.
3. Use these values to calculate the length of the edges a and b.
4. Treat A and B as a composite. Calculate the average distances between AB and each of the other sequences. Create a new distance table.
5. Identify next pair of related sequences and begin as with step 1.
6. Subtract extended branch lengths to calculate lengths of intermediate branches.
7. Repeat the entire process with all possible pairs of sequences.
8. Calculate predicted distances between each pair of sequences for each tree to find the best tree.

23.16.3 Neighbour-joining algorithm

The neighbour-joining method is very similar to the Fitch-Margoliash method. The sequences that should be joined are chosen to give the best least-squares estimates of the branch lengths that most closely reflect the actual distances between the sequences. The neighbour-joining method begins by creating a star topology in which no neighbours are joined (Fig. 23.8).

23.16.4 Unweighted pair group method with arithmetic mean

Works by clustering the sequences, starting with more similar sequences and working towards more distant sequences. The process assembles a tree upwards, with each node being added above the others and the edge lengths being determined by the difference in the heights of the nodes.

23.16.5 Difficulties with phylogenetic analysis

Phylogenetic analysis would be easier if evolution occurred in a vertical fashion. However, horizontal or lateral transfer of genetic material (for instance through viruses) occurs, which makes it difficult to determine the phylogenetic origin of some evolutionary events.

Selective pressure: If a gene is under selective pressure in different organisms, it can be rapidly evolving. Such an evolution can mask earlier changes that had occurred phylogenetically. In addition, different regions of a genome are under different pressures and therefore different sites within two comparative sequences may be evolving at different rates.

Rearrangement of genetic material: Rearrangements of genetic material can also lead to false conclusions with phylogenetic analysis, especially if two sequences of different evolutionary origins are placed next to each other.

Gene duplication: Gene duplication events also cause problems with phylogenetic analysis, since the duplicated genes can evolve along separate pathways, leading to different functions.

PAUP (maximum parsimony)

MacClade (maximum parsimony)

PHYLIP—(distance—neighbour joining)

CLUSTALW—distance-based tree

Consider the following list of Globin sequences:

>gamma_A
MGHFTEEDKATITSLWGKVNVEDAGGETLGRLLVVYPWTQRFFDSFGNLSSASAIMGNPKV
KAHGKKVLTSLGDAIKHLDDLKGTFAQLSELHCDKLHVDPENFKLLGNVLVIVIAIHFGKEF
TPEVQASWQKMVTAVAS ALSSRYH
>alfa
VLSPADKTNVKAAWGKVGAHAGEYGAEALERMFLSFPTTKTYFPHFDLSHGSAQVKGHG
KKVADALTNAVAHVDDMPNALSALSDLHAHKLRVDPVNFKLLSHCLLVTLAAHLPAEFTPA
VHASLDKFLASVSTVLTSKYR
>beta
VHLTPEEKSAVTALWGKVNVDEVGGEALGRLLVVYPWTQRFFESFGDLSTPDAVMGNPKV
KAHGKKVLGAFSDGLAHLDNLKGTFATLSELHCDKLHVDPENFRLLGNVLVCVLAHHFGK
EFTPPVQAAYQKVVAGVANALAHKYH

>delta
VHLTPEEKTAVNALWGKVNVDAVGGEALGRLLVVYPWTQRFFESFGDLSSPDAVMGNPKV
KAHGKKVLGAFSDGLAHLDNLKGTFSQLSELHCDKLHVDPENFRLLGNVLVCVLARNFGK
EFTPQMQAAYQKVVAGVANALAHKYH
>epsilon
VHFTAEEKAAVTSLWSKMNVEEAGGEALGRLLVVYPWTQRFFDSFGNLSSPSAILGNPKV
KAHGKKVLTSFGDAIKNMDNLKPAFAKLSELHCDKLHVDPENFKLLGNVMVIILATHFGK
EFTPEVQAAWQKLVSAVAIALAHKYH
>gamma_G
MGHFTEEDKATITSLWGKVNVEDAGGETLGRLLWYPWIQRFFDSFGNLSSASAIMGNPKV
KAHGKKVLTSLGDAIKHLDDLKGTFAQLSELHCDKLHVDPENFKLLGNVLVTVIAIHFGKE
FTPEVQASWQKMVTGVASALSSRYH
>myoglobin
MGLSDGEWQLVLNVWGKVEADIPGHGQEVLIRLFKGHPETLEKFDKFKHLKSEDEMKASE
DLKKHGATVLTALGGILKKKGHHEAEIKPLAQSHATKHKIPVKYLEFISECIIQVLQSKHPGD
FGADAQGAMNKALELFRKDMASNYKELGFQG
>tetal
ALSAEDRALVRALWKKLGSNVGVYTTEALERTFLAFPATKTYFSHLDLSPGSSQVRAHGQ
KVADALSLAVERLDDLPHALSALSHLHACQLRVDPASFQLLGHCLLVTLARHYPGDFSPA
LQASLDKFLSHVISALVSEYR
>zeta
SLTKTERTIIVSMWAKISTQADTIGTETLERLFLSHPQTKTYFPHFDLHPGSAQLRAHGS
KVVAAVGDAVKSIDDIGGALSKLSELHAYILRVDPVNFKLLSHCLLVTLAARFPADFTAE
AHAAWDKFLSVVSSVLTEKYR
    Create a phylogeny from these.

23.17   Phylogenetic tree

23.17.1 Methods to build a tree

1. Genetic distances.
2. Phylogenetic analysis programmes and where to get them.
3. Our interface.
4. Common mistakes in making trees.
5. What to do when your tree looks funny.

    First of all, it is important to realise that all tree-building methods assume that the alignment is correct; errors in the alignment can lead to a very misleading tree. Once the alignment is optimal, there are many different methods to create phylogenies. Roughly, the methods can be divided into distance-based and character-based methods. Character-based methods use the individual substitutions among the sequences to determine the most likely ancestral relationships, while distance-based methods first calculate the overall distance between all pairs of sequences and then calculate a tree based on those distances. Maximum parsimony (MP) and maximum likelihood (ML) are the most important character based methods. In most comparative studies, ML seems to be the method that yields the best trees. The most important drawback of ML is that it is very computationally intensive; it is almost unusable with more than a few dozen sequences. MP is much faster than ML, but still slow compared to most distance-based methods. Both these programmes calculate large numbers of trees and compare them by either the likelihood or the parsimony score. There are programmes available which speed up the process considerably, such as FastDNAml but for large numbers of sequences the computational burden is still a major hurdle.

The distance-based methods are generally much faster than the character-based ones. In this group, neighbour joining (NJ) is by far the most popular method. It is very fast and generally quite good, although there are conditions under which it systematically produces a wrong tree (bias). NJ is not the only biased method: in fact, under the right conditions (mostly when evolutionary rates are different for different sites, which is frequently the case for HIV) any method is biased, i.e., systematically produces the wrong tree. The importance of the bias for the everyday tree is a matter of fierce debate.

Weighbor (short for 'weighted neighbour-joining') is a new method by Bruno, Socci and Halpern, which gives less weight to the longer distances in the distance matrix. The resulting trees are less sensitive to specific biases than NJ and MP and negative branch lengths are avoided. The method is much faster than ML and usually faster than MP, but much slower than NJ.

To some extent the choice between methods can be based on the purpose of the tree. For subtyping sequences, an NJ tree based on a matrix of genetic distances is generally good enough. It is not vital that the tree is correct in every branch, only that the sequence of interest is clustered with the right subtype. Almost any method will solve this problem correctly.

However, when more detailed information on the evolutionary relationships is important, for instance in forensic analyses or when studying rates of evolution or trying to resolve key relationships (for example, short domains and potential recombination), more realistic models of evolution and unbiased tree reconstruction methods should be used.

Genetic distances

The simplest way to calculate a distance between two sequences is to count the number of differences. This measure is often called the hamming distance. It reflects the difference between sequences, but ignores their evolutionary relationship. The most intuitive way to show how this can be misleading to think of superimposed or reversed mutations: a nucleotide in a sequence, say an A, gets mutated and becomes a G. This results in a difference of 1. Many replication rounds later the same site gets mutated again. If it becomes an A, the difference now goes down, even though the genetic distance (the number of mutational events) has gone up. If it becomes a T or a C, the differences doesn't change, but the evolutionary distance should increase, because there have now been two evolutionary events. The oldest attempt to adjust the number of differences between two sequences for the chances of a parallel or back mutation was designed by Jukes and Cantor. They proposed the corrective formula $D = -3/4 \ln[1-(4p/3)]$, where, D = Distance, ln = Leen, P = Base frequency. First, the effect of the correction increases with the difference between the sequences and is negligible with very homologous sequences. Second, the effect of saturation can be seen beyond a certain number of differences (75%), it becomes impossible to tell what the true genetic distance is. Since this formula was first proposed in the 1980s, many different models have been proposed to accurately estimate the underlying evolutionary distance from the observed number of differences between the sequences, taking into account new knowledge about the behaviour of DNA, such as different transition/transversion rates, different base frequencies and nonuniform substitution rates between sites.

Not all of these sophisticated methods are easily available (Dnadist, for example, only offers Jukes-Cantor, Kimura 2-parametre, Jin-Nei and the F84 model Phylip uses for its maximum likelihood trees; Mega also offers Tamura and Tamura-Nei and a set of Gamma distribution-based distances).

A relatively new development is the use of models that incorporate variable/ evolutionary rates across sites. This is an important extension of the existing models, especially for HIV, which is well known to show dramatically different evolutionary rates (e.g., under the influence of immune escape) over the length of the genome. Gary Olsen has a programme called DNA rates, which can estimate the rates of evolution in different sites, given an initial tree. The number of categories for the rates can be specified by the user.

Using the DNA rates programme can quite dramatically increase the quality of the resulting trees. Many simulations show that the importance of using a realistic substitution model for estimating the genetic distance depends very much on the divergence of the sequences. If the expected number of substitutions per site is small (below 0.2), the resulting tree will not change much when different substitution models are used. If the sequences are highly diverged, however and especially when the number of substitutions per site rises above 1, the differences become marked, the choice of the substitution model is very important.

Tree-building programmes

Probably the most widely used programme suite is Felsenstein's PHYLIP. It offers a large array of methods, including ML, MP and NJ. The choice of genetic distance is fairly limited and learning to use the programmes requires some time investment, but their versatility makes them very popular. Most allow input of user trees, jumbling (randomising input orders) and out group designation.

MEGA is a nifty little programme for DOS/Windows. It offers NJ and MP (although in a quirky implementation) and the sophisticated Tamura-Nei genetic distance estimation.

MEGA is able to build trees on the basis of silent or nonsilent mutations only and has a fast bootstrapping option built in. Downsides are that the programme is fairly unstable and has memory problems with large sets of sequences and the printing of the trees often gives problems.

PHYLOWIN is a UNIX/Linux based programme with a very nice graphical user interface. It does ML, MP and NJ, allows selection of sequences (rows) and position (columns) and bootstrapping. It allows the User to define subsets of sequences and positions and save them. It is available free of charge for academic users. FastDNAml computes fast(er) ML trees. It is not very easy to use, but is often the only option if one wants to calculate ML trees for more than 20 sequences. It uses input file or command line parametres to specify the trees. The C source code for the programme and an executable for Powermac are available by ftp.

PAUP is a very versatile programme that does maximum likelihood tree building in addition to parsimony and allows incorporation of variable rates per site. It is presently distributed as a beta version. The Mac version is entirely menu-driven, but the user interface is very clumsy. There are no user interfaces for the Windows and Unix versions.

23.17.3 HIV-WEB treemaker interface

This interface is just that: it interfaces between the user and the phylogenetic programmes Dnadist, Neighbour and Drawtree/Drawgram from Joe Felsenstein's PHYLIP suite. The interface can only be used to make Neighbour-Joining trees, which may not be optimal for all circumstances, but usually form a good starting point for more sophisticated analysis, and as mentioned above, if the inferences made from the tree need not be very exact, for example for subtyping a small region or for a quick contamination check, this tree can suffice. Please note that the interface does not do bootstrapping.

The interface takes a sequence alignment in several formats, allows the adjustment of a few parametres (transition/transversion ratio, outgroup and the shape of the tree). It uses the F84 genetic distance estimate (i.e., the option called 'ML' in Dnadist).

23.17.4 Common mistakes in making a tree

Presenting the output from PHYLIPs consense programme as the final tree

This tree gives only a branching order. The 'branchlengths' are not true branch lengths, but rather reflect the per cent bootstrap values. For this reason, the Treefile that consense produces does not contain branch lengths and when printed, all branches in the tree have the same length. Remedy: If you want to include bootstrap values in a tree created with PHYLIP, the simplest method is to simply paste the values into the nonbootstrapped tree with valid branchlengths. The better way is to use the tree you get from the Consense programme as input for another run of the tree-building programme to have the branch lengths estimated for that particular tree. This also solves the (infrequent) problem where the topology of the consensus tree doesn't exactly match the one of the original tree.

Visible alignment errors and/or unrecognised hypermutation

When one sequence protrudes far out beyond all the others and there is no inherent reason for it to be so different, further inspection is needed. Frequently the cause is an alignment error or a hypermutated sequence.

Remedy:
1. Visually check the alignment of your sequences. There are many alignment editors available that make it very easy to do this.
2. Check your sequence for hypermutation. Hypermutation, a relatively common phenomenon in HIV, means a very high incidence of G - > A mutations, usually resulting in a nonviable sequence. An interface that was designed to detect hypermutation: HYPERMUT.

Unrecognised recombination

When an isolate branches off close to the root between two subtypes, especially if it has a long branch or if it is not from a particularly old isolate, there is a chance that it is a recombinant. In this case the MAL isolate, an A/D/I recombinant. Similarly, when a sequence branches off between two clusters from one patient isolate, it can be a within-patient recombinant. In this case it can be the result of a real recombination event or a PCR artifact.

Remedy: If you suspect your sequence may be a recombinant, there are a multitude of ways to look at this more closely. On this we have RIP, which produces an alignment and an easy-to-read plot that shows the similarity of your sequence to a set of reference sequences over the entire length of the sequence. If there are major changes in what the most similar sequence is over the length of your sequence, this suggests recombination. There are many more methods to detect recombination.

Assuming a molecular clock

The graphical representation of this assumption is that all branches end on one vertical line, representing the present day. This assumption is not realistic for HIV and it is very uncommon for other organisms. The most commonly used method that produces these trees is UPGMA, the Kitsch (which explicitly assumes a molecular clock) programme from the PHYLIP suite also results in this type of tree.

Remedy: Use a different tree reconstruction method. Neighbour joining, maximum parsimony and maximum likelihood all produce trees that do not assume a strict molecular clock.

23.17.5 What to do when your tree looks funny

If none of these errors describe your situation, but you still think your tree is off, there are a few things you can do.

1. Try a different method or a different distance estimate. In rare cases (when there are many equivalent trees) even the input order of the sequence can make a difference, use the Jumble option provided in DNAML or rearrange your input file.

2. Try using a different programme that uses the same tree reconstruction method and compare the results.

3. Use a different outgroup: although the outgroup does not affect the structure of the tree, it can sometimes make it easier to interpret.

4. Split your sequences in half and see if the resulting trees are different. This suggests either recombination or dramatic evolutionary rate differences. In some cases (an example is the Rev-responsive element or RRE in HIV) two adjacent regions can be under such different constraint that they evolve very differently; building a tree from a sequence that spans both regions can give confusing results.

23.17.6 Tree reliability

By far the most popular test for trees is the bootstrap. Contrary to what many people think, this is not a test of how accurate your tree is; it only gives information about the stability of the tree topology (the branching order) and it helps assess whether the sequence data is adequate to validate the topology. The bootstrap randomly resamples columns from your alignment, so that some positions will not be used and others will be used more than once and builds a new tree from this dataset. This is done as many times as you specify. The bootstrap value is a count or percentage of how often each branch was present in exactly the same topology in all the resampled trees, so it gives an impression of how much the tree topology could change if, for example, you'd reconstruct it using a different gene. There are many rules of thumb about how to interpret the bootstrap. It is known to be a conservative measure, so a bootstrap of 95 per cent gives more than 95 per cent confidence in that branch. The number of 70 per cent is often cited as a cut-off for a 'reliable' branch.

Epigenetics is defined as heritable changes in gene activity and expression that occur without alteration in DNA sequence. It is known these non-genetic alternations are tightly regulated by two major epigenetic modifications: chemical modifications to the cytosine residues of DNA (DNA methylation) and histone proteins associated with DNA (histone modifications). Functionally, the patterns of epigenetic modifications can serve as epigenetic markers to represent gene activity and expression as well as chromatin state.

Epigenetic mechanisms regulate gene function in a heritable manner, but do so without modulating the DNA sequence of the affected gene. Many different genetic functions are influenced by epigenetic mechanisms in various species. These include regulation of gene expression, DNA modification and restriction, genomic imprinting, X-chromosome inactivation, paramutation, position effect variegation, mating type, cell determination, transposable elements and mutator and suppressor genes.

Nuclear DNA acts as the repository of genetic information in eukaryotic cells. In mammals, and many other animal species, a complete representation of the genome is maintained in essentially every nucleated cell. However, only a subset of this collection of genes is expressed in any particular cell type. Thus, it is not the presence of specific genes, but rather the expression of specific genes, that leads to the unique identity and function of any particular cell. For protein-encoding genes, two primary steps are involved in gene expression: transcription of DNA into RNA and translation of that RNA into a polypeptide. This affords two levels of regulation of gene expression: transcriptional regulation and translational (or post-transcriptional) regulation. For tissue-specific genes (those expressed in only a subset of tissues or in a single tissue or cell type), regulation is primarily manifest at the transcriptional level. Extensive studies of this process have revealed a consensus mechanism whereby the promoter region, typically located at the 5'-end of the gene, acts to bind specific proteins called transcription factors which, in turn, attract (or prevent) binding of the RNA polymerase that is required to initiate transcription. Binding of transcription factors to specific gene promoters and to specific sites within those promoters is regulated by the ability of a DNA binding domain within each protein factor to recognise a unique three-dimensional structure of double stranded DNA.

This unique structure is imparted by a specific nucleotide sequence, typically 5–15 base pairs (bp) in length. Thus, this mechanism does rely on the DNA sequence and is therefore not a truly epigenetic mechanism. However, because these transcription factors can be either ubiquitous or tissue-specific, and can either promote or inhibit transcription, this mechanism can modulate tissue, cell-type or developmental-stage specificity of transcription, as well as controlling the relative level (or frequency) of transcription.

Nevertheless, protein-DNA interactions between transcription factors and promoter sequences, respectively, are not the only mechanism by which gene expression is regulated in eukaryotic cells.

In mammals there are several examples in which genes are regulated by mechanisms other than transcription factors. For example, in female somatic cells, genes on the active X-chromosome are transcribed, whereas homologous genes on the inactive X-chromosome remain transcriptionally silenced. This is despite the fact that both the active and inactive copies of these genes share identical nucleotide sequences and reside within the same nucleus.

Thus the presence of identical promoter sequences and cognate transcription factors alone does not ensure identical regulation of genes. Similarly, in mammals, the phenomenon of genomic imprinting results in the expression of only one of the two copies of a particular gene within a single diploid cell. In this case the choice of which allele is expressed is dictated by the parental origin of that allele. However, the mechanism that regulates such monoallelic expression cannot be based solely on transcription factors and promoter sequences, because the former are present throughout the nucleus in which both alleles reside, and the latter are often identical on both alleles.

The unavoidable conclusion from these observations is that there must be additional mechanisms by which gene expression is regulated in eukaryotic cells, and these mechanisms must function in a manner that does not depend on differences in nucleotide sequence or the cell-type specific presence or absence of transcription factors. Yet, as exemplified by the examples noted earlier for X-chromosome inactivation and genomic imprinting, these mechanisms must function in a heritable manner, such that the same alleles remain expressed or silenced, even after replication of the DNA and division of one cell to produce two daughter cells.

We now know that there are multiple mechanisms that meet the criteria of epigenetic mechanisms in that they regulate gene expression in a heritable manner that does not rely on differences in DNA sequence. Examples of mechanisms that either have been shown to operate in this manner or have the potential to operate in this manner include: DNA methylation, chromatin structure and/or composition, DNA loop domains and association with the nuclear matrix and DNA replication timing.

23.18 DNA methylation

DNA methylation is a biochemical process where a methyl group is added to the cytosine or adenine DNA nucleotides. The rate of cytosine DNA methylation differs strongly between species, e.g., absolute quantification by mass spectrometry revealed 14% of cytosines methylated in Arabidopsis thaliana, 8% in Mus musculus, 2.3% in *Escherichia coli*, 0.03% in *Drosophila*, and virtually none (< 0.0002%) in yeast species. DNA methylation may stably alter the expression of genes in cells as cells divide and differentiate from embryonic stem cells into specific tissues.

The resulting change is normally permanent and unidirectional, preventing a cell from reverting to a stem cell or converting into a different cell type. DNA methylation is typically removed during zygote formation and re-established through successive cell divisions during development. However, the latest research shows that hydroxylation of methyl groups occurs rather than complete removal of methyl groups in the zygote. Some methylation modifications that regulate gene expression are heritable and cause genomic imprinting.

DNA methylation suppresses the expression of endogenous retroviral genes and other harmful stretches of DNA that have been incorporated into the host genome over time. DNA methylation also forms the basis of chromatin structure, which enables a single cell to grow into multiple organs or perform multiple functions. DNA methylation also plays a crucial role in the development of nearly all types of cancer.

DNA methylation at the 5 position of cytosine has the specific effect of reducing gene expression and has been found in every vertebrate examined. In adult somatic cells (cells in the body, not used for reproduction), DNA methylation typically occurs in a CpG dinucleotide context, non-CpG methylation is prevalent in embryonic stem cells, and has also been indicated in neural development.

23.18.1 In mammals

DNA methylation is essential for normal development and is associated with a number of key processes including genomic imprinting, X-chromosome inactivation, suppression of repetitive elements, and carcinogenesis.

Between 60% and 90% of all CpGs are methylated in mammals. Methylated C residues spontaneously deaminate to form T residues over time; hence CpG dinucleotides steadily deaminate to TpG dinucleotides, which is evidenced by the under-representation of CpG dinucleotides in the human genome (they occur at only 21% of the expected frequency). (On the other hand, spontaneous deamination of unmethylated C residues gives rise to U residues, a change that is quickly recognised and repaired by the cell.)

Unmethylated CpGs are often grouped in clusters called CpG islands, which are present in the 5′ regulatory regions of many genes. In many disease processes, such as cancer, gene promoter CpG islands acquire abnormal hypermethylation, which results in transcriptional silencing that can be inherited by daughter cells following cell division. Alterations of DNA methylation have been recognised as an important component of cancer development.

Hypomethylation, in general, arises earlier and is linked to chromosomal instability and loss of imprinting, whereas hypermethylation is associated with promoters and can arise secondary to gene (oncogene suppressor) silencing, but might be a target for epigenetic therapy.

DNA methylation may affect the transcription of genes in two ways. First, the methylation of DNA itself may physically impede the binding of transcriptional proteins to the gene, and second, and likely more important, methylated DNA may be bound by proteins known as methyl-CpG-binding domain proteins (MBDs). MBD proteins then recruit additional proteins to the locus, such as histone deacetylases and other chromatin remodelling proteins that can modify histones, thereby forming compact, inactive chromatin, termed heterochromatin. This link between DNA methylation and chromatin structure is very important. In particular, loss of methyl-CpG-binding protein 2 (MeCP2) has been implicated in Rett syndrome; and methyl-CpG-binding domain protein 2 (MBD2) mediates the transcriptional silencing of hypermethylated genes in cancer.

Research has suggested that long-term memory storage in humans may be regulated by DNA methylation. DNA methylation levels can be used to estimate age, forming an accurate biological clock in humans and chimpanzees.

In cancer

DNA methylation is an important regulator of gene transcription and a large body of evidence has demonstrated that genes with high levels of 5-methyl-cytosine in their promoter region are transcriptionally silent, and that DNA methylation gradually accumulates upon long-term gene silencing. DNA methylation is essential during embryonic development, and in somatic cells, patterns of DNA methylation are generally transmitted to daughter cells with a high fidelity. Aberrant DNA methylation patterns–hypermethylation and hypomethylation compared to normal tissue–have been associated with a large number of human malignancies. Hypermethylation typically occurs at CpG islands in the promoter region and is associated with gene inactivation. A lower level of leukocyte DNA methylation is associated with many types of cancer. Global hypomethylation has also been implicated in the development and progression of cancer through different mechanisms. Typically, there is hyper-methylation of tumor suppressor genes and hypomethylation of oncogenes.

In atherosclerosis

Epigenetic modifications such as DNA methylation have been implicated in cardiovascular disease, including atherosclerosis. In animal models of atherosclerosis, vascular tissue as well as blood cells such as mononuclear blood cells exhibit global hypomethylation with gene-specific areas of hyper-methylation. DNA methylation polymorphisms may be used as an early biomarker of atherosclerosis since they are present before lesions are observed, which may provide an early tool for detection and risk prevention.

Two of the cell types targeted for DNA methylation polymorphisms are monocytes and lymphocytes, which experience an overall hypomethylation. One proposed mechanism behind this global hypomethylation is elevated homocysteine levels causing hyperhomocysteinemia, a known risk factor for cardiovascular disease. High plasma levels of homocysteine inhibit DNA methyltransferases, which causes hypomethylation. Hypomethylation of DNA affects gene that alter smooth muscle cell proliferation, cause endothelial cell dysfunction, and increase inflammatory mediators, all of which are critical in forming atherosclerotic lesions. High levels of homocysteine also result in hypermethylation of CpG islands in the promoter region of the estrogen receptor alpha (ERα) gene, causing its down regulation. ERα protects against atherosclerosis due to its action as a growth suppressor, causing the smooth muscle cells to remain in a quiescent state. Hypermethylation of the ERα promoter thus allows intimal smooth muscle cells to proliferate excessively and contribute to the development of the atherosclerotic lesion.

Another gene that experiences a change in methylation status in atherosclerosis is the monocarboxylate transporter (MCT3), which produces a protein responsible for the transport of lactate and other ketone bodies out of many cell types, including vascular smooth muscle cells. In atherosclerosis patients, there is an increase in methylation of the CpG islands in exon 2, which decreases MCT3 protein expression. The down regulation of MCT3 impairs lactate transport, and significantly increases smooth muscle cell proliferation, which further contributes to the atherosclerotic lesion. An ex vivo experiment using the demethylating agent Decitabine (5-aza-2 -deoxycytidine) was shown to induce MCT3 expression in a dose dependant manner, as all hypermethylated sites in the exon 2 CpG island became demethylated after treatment. This may serve as a novel therapeutic agent to treat atherosclerosis, although no human studies have been conducted thus far.

In ageing

A longitudinal study of twin children, showed that between the ages of 5 and 10 there was divergence of methylation patterns due to environmental rather than genetic influences. There is a global loss of DNA methylation during aging. But some genes become hypermethylated with age, including genes for the estrogen receptor, p16, and insulin-like growth factor 2. Biological clocks such as an epigenetic clock, are promising biomarkers of ageing.

In exercise

High intensity exercise has been shown to result in reduced DNA methylation in skeletal muscle. Promoter methylation of PGC-1a and PDK4 were immediately reduced after high intensity exercise, whereas PPAR-a methylation was not reduced until three hours after exercise. By contrast, six months of exercise in previously sedentary middle-age men resulted in increased methylation in adipose tissue. One study showed a possible increase in global genomic DNA methylation of white blood cells with more physical activity in non-Hispanics.

23.18.2 DNA methyltransferases

In mammalian cells, DNA methylation occurs mainly at the C5 position of CpG dinucleotides and is carried out by two general classes of enzymatic activities– maintenance methylation and *de novo* methylation.

Maintenance methylation activity is necessary to preserve DNA methylation after every cellular DNA replication cycle. Without the DNA methyltransferase (DNMT), the replication machinery itself would produce daughter strands that are unmethylated and, over time, would lead to passive demethylation. DNMT1 is the proposed maintenance methyltransferase that is responsible for copying DNA methylation patterns to the daughter strands during DNA replication. Mouse models with both copies of DNMT1 deleted are embryonic lethal at approximately day 9, due to the requirement of DNMT1 activity for development in mammalian cells.

It is thought that DNMT3a and DNMT3b are the *de novo* methyltransferases that set up DNA methylation patterns early in development. DNMT3L is a protein that is homologous to the other DNMT3s but has no catalytic activity. Instead, DNMT3L assists the de novo methyltransferases by increasing their ability to bind to DNA and stimulating their activity. Finally, DNMT2 (TRDMT1) has been identified as a DNA methyltransferase homolog, containing all 10 sequence motifs common to all DNA methyltransferases; however, DNMT2 (TRDMT1) does not methylate DNA but instead methylates cytosine-38 in the anticodon loop of aspartic acid transfer RNA.

Since many tumor suppressor genes are silenced by DNA methylation during carcinogenesis, there have been attempts to re-express these genes by inhibiting the DNMTs. 5-Aza-2′-deoxycytidine (decitabine) is a nucleoside analog that inhibits DNMTs by trapping them in a covalent complex on DNA by preventing the β-elimination step of catalysis, thus resulting in the enzymes degradation.

However, for decitabine to be active, it must be incorporated into the genome of the cell, which can cause mutations in the daughter cells if the cell does not die. In addition, decitabine is toxic to the bone marrow, which limits the size of its therapeutic window. These pitfalls have led to the development of antisense RNA therapies that target the DNMTs by degrading their mRNAs and preventing their translation. However, it is currently unclear whether targeting DNMT1 alone is sufficient to reactivate tumor suppressor genes silenced by DNA methylation.

In plants

Significant progress has been made in understanding DNA methylation in the model plant Arabidopsis thaliana. DNA methylation in plants differs from that of mammals: while DNA methylation in mammals mainly occurs on the cytosine nucleotide in a CpG site, in plants the cytosine can be methylated at CpG, CpHpG, and CpHpH sites, where H represents any nucleotide but guanine. Overall, Arabidopsis DNA is highly methylated, i.e. an exemplary mass spectrometry analysis estimated 14% of cytosines to be modified.

The principal Arabidopsis DNA methyltransferase enzymes, which transfer and covalently attach methyl groups onto DNA, are DRM2, MET1, and CMT3. Both the DRM2 and MET1 proteins share significant homology to the mammalian methyl-transferases DNMT3 and DNMT1, respectively, whereas the CMT3 protein is unique to the plant kingdom. There are currently two classes of DNA methyltransferases: (i) the *de novo* class, or enzymes that create new methylation marks on the DNA and (ii) a maintenance class that recognises the methylation marks on the parental strand of DNA and transfers new methylation to the daughters strands after DNA replication. DRM2 is the only enzyme that has been implicated as a de novo DNA methyltransferase. DRM2 has also been shown, along with MET1 and CMT3 to be involved in maintaining methylation marks through DNA replication. Other DNA methyltransferases are expressed in plants but have no known function.

It is not clear how the cell determines the locations of de novo DNA methylation, but evidence suggests that, for many (though not all) locations, RNA-directed DNA methylation (RdDM) is involved. In RdDM, specific RNA transcripts are produced from a genomic DNA template, and this RNA forms secondary structures called double-stranded RNA molecules. The double-stranded RNAs, through either the small interfering RNA (siRNA) or microRNA (miRNA) pathways direct *de novo* DNA methylation of the original genomic location that produced the RNA. This sort of mechanism is thought to be important in cellular defense against RNA viruses and/or transposons, both of which often form a double-stranded RNA that can be mutagenic to the host genome. By methylating their genomic locations, through an as yet poorly understood mechanism, they are shut off and are no longer active in the cell, protecting the genome from their mutagenic effect.

In fungi

Many fungi have low levels (0.1 to 0.5%) of cytosine methylation, whereas other fungi have as much as 5% of the genome methylated. This value seems to vary both among species and among isolates of the same species. There is also evidence that DNA methylation may be involved in state-specific control of gene expression in fungi. However, at a detection limit of 250 attomoles by using ultra-high sensitive mass spectrometry DNA methylation was not confirmed in single cellular yeast species such as *Saccharomyces cerevisiae* or *Schizosaccharomyces pombe*, indicating that yeasts do not possess this DNA modification. Although brewers' yeast (*Saccharomyces*) and fission yeast (*Schizosaccharomyces*) have no detectable DNA methylation, the model filamentous fungus Neurospora crassa has a well-characterised methylation system. Several genes control methylation in Neurospora and mutation of the DNA methyl transferase, dim-2, eliminates all DNA methylation but does not affect growth or sexual reproduction. While the Neurospora genome has very little repeated DNA, half of the methylation occurs in repeated DNA including transposon relics and centromeric DNA. The ability to evaluate other important phenomena in a DNA methylase-deficient genetic background makes Neurospora an important system in which to study DNA methylation.

In insects

DNA methylation is debated in insects, *Drosophila melanogaster* for instance seems to possess a very low level of DNA methylation only that is however too low to be studied by methods such as bisulphite sequencing. Takayama and others developed a sensitive method that allowed to find that the fly genome DNA sequence patterns that associate with methylation are very different from the patterns seen in humans, or in other animal or plant species to date. Genome methylation in *D. melanogaster* was found at specific short motifs (concentrated in specific 5-base sequence motifs that are CA- and CT-rich but depleted of guanine) and is independent of DNMT2 activity.

In bacteria

Adenine or cytosine methylation is part of the restriction modification system of many bacteria, in which specific DNA sequences are methylated periodically throughout the genome. A methylase is the enzyme that recognises a specific sequence and methylates one of the bases in or near that sequence. Foreign DNAs (which are not methylated in this manner) that are introduced into the cell are degraded by sequence-specific restriction enzymes and cleaved.

Bacterial genomic DNA is not recognised by these restriction enzymes. The methylation of native DNA acts as a sort of primitive immune system, allowing the bacteria to protect themselves from infection by bacteriophage.

E. coli DNA adenine methyltransferase (Dam) is an enzyme of ~32 kDa that does not belong to a restriction/modification system. The target recognition sequence for *E. coli* Dam is GATC, as the methylation occurs at the N6 position of the adenine in this sequence (G meATC). The three base pairs flanking each side of this site also influence DNA–Dam binding. Dam plays several key roles in bacterial processes, including mismatch repair, the timing of DNA replication, and gene expression. As a result of DNA replication, the status of GATC sites in the *E. coli* genome changes from fully methylated to hemi-methylated. This is because adenine introduced into the new DNA strand is unmethylated. Re-methylation occurs within two to four seconds, during which time replication errors in the new strand are repaired. Methylation, or its absence, is the marker that allows the repair apparatus of the cell to differentiate between the template and nascent strands. It has been shown that altering Dam activity in bacteria results in increased spontaneous mutation rate. Bacterial viability is compromised in dam mutants that also lack certain other DNA repair enzymes, providing further evidence for the role of Dam in DNA repair.

One region of the DNA that keeps its hemimethylated status for longer is the origin of replication, which has an abundance of GATC sites. This is central to the bacterial mechanism for timing DNA replication. SeqA binds to the origin of replication, sequestering it and thus preventing methylation. Because hemimethylated origins of replication are inactive, this mechanism limits DNA replication to once per cell cycle.

Expression of certain genes, for example those coding for pilus expression in *E. coli*, is regulated by the methylation of GATC sites in the promoter region of the gene operon. The cells environmental conditions just after DNA replication determine whether Dam is blocked from methylating a region proximal to or distal from the promoter region. Once the pattern of methylation has been created, the pilus gene transcription is locked in the on or off position until the DNA is again replicated. In *E. coli*, these pilus operons have important roles in virulence in urinary tract infections. It has been proposed that inhibitors of Dam may function as antibiotics. On the other hand, DNA cytosine methylase targets CCAGG and CCTGG sites to methylate cytosine at the C5 position (C meC(A/T) GG). The other methylase enzyme, EcoKI, causes methylation of adenines in the sequences AAC(N$_6$)GTGC and GCAC(N$_6$)GTT.

23.19 Chromatin

Chromatin is the combination or complex of DNA and proteins that make up the contents of the nucleus of a cell. The primary functions of chromatin are:

(i) to package DNA into a smaller volume to fit in the cell, (ii) to strengthen the DNA to allow mitosis, (iii) to prevent DNA damage and (iv) to control gene expression and DNA replication. The primary protein components of chromatin are histones that compact the DNA. Chromatin is only found in eukaryotic cells, (a cell with a defined nucleus). Prokaryotic cells have a very different organisation of their DNA, which is referred to as a genophore (a chromosome without chromatin). The structure of chromatin depends on several factors. The overall structure depends on the stage of the cell cycle. During interphase, the chromatin is structurally loose to allow access to RNA and DNA polymerases that transcribe and replicate the DNA.

The local structure of chromatin during interphase depends on the genes present on the DNA: DNA coding genes that are actively transcribed ('turned on') are more loosely packaged and are found associated with RNA polymerases (referred to as euchromatin) while DNA coding inactive genes ('turned off') are found associated with structural proteins and are more tightly packaged (heterochromatin). Epigenetic chemical modification of the structural proteins in chromatin also alter the local chromatin structure, in particular chemical modifications of histone proteins by methylation and acetylation. As the cell prepares to divide, i.e. enters mitosis or meiosis, the chromatin packages more tightly to facilitate segregation of the chromosomes during anaphase. During this stage of the cell cycle this makes the individual chromosomes in many cells visible by optical microscope.

In general terms, there are three levels of chromatin organisation:

1. DNA wraps around histone proteins forming nucleosomes; the 'beads on a string' structure (euchromatin).

2. Multiple histones wrap into a 30 nm fibre consisting of nucleosome arrays in their most compact form (heterochromatin). (Definitively established to exist *in vitro*, the 30-nanometer fibre was not seen in recent X-ray studies of human mitotic chromosomes.)

3. Higher-level DNA packaging of the 30 nm fibre into the metaphase chromosome (during mitosis and meiosis).

There are, however, many cells that do not follow this organisation. For example, spermatozoa and avian red blood cells have more tightly packed chromatin than most eukaryotic cells, and trypanosomatid protozoa do not condense their chromatin into visible chromosomes for mitosis.

23.19.1 During interphase

The structure of chromatin during interphase is optimised to allow easy access of transcription and DNA repair factors to the DNA while compacting the DNA into the nucleus. The structure varies depending on the access required to the DNA. Genes that require regular access by RNA polymerase require the looser structure provided by euchromatin.

23.19.2 Change in structure

Chromatin undergoes various forms of change in its structure. Histone proteins, the foundation blocks of chromatin, are modified by various post-translational modification to alter DNA packing. Acetylation results in the loosening of chromatin and lends itself to replication and transcription. When certain residues are methylated, they hold DNA together strongly and restrict access to various enzymes. A recent study showed that there is a bivalent structure present in the chromatin: methylated lysine residues at location 4 and 27 on histone 3. It is thought that this may be involved in development, there is more methylation of lysine 27 in embryonic cells than in differentiated cells, whereas lysine 4 methylation positively regulates transcription by recruiting nucleosome remodelling enzymes and histone acetylases. Polycomb-group proteins play a role in regulating genes through modulation of chromatin structure.

23.19.3 DNA structure

The vast majority of DNA within the cell is the normal DNA structure. However, in nature, DNA can form three structures, A-, B-, and Z-DNA. A and B chromosomes are very similar, forming right-handed helices, whereas Z-DNA is a more unusual left-handed helix with a zig-zag phosphate backbone. Z-DNA is thought to play a specific role in chromatin structure and transcription because of the properties of the junction between B- and Z-DNA. At the junction of B- and Z-DNA, one pair of bases is flipped out from normal bonding. These play a dual role of a site of recognition by many proteins and as a sink for torsional stress from RNA polymerase or nucleosome binding.

Nucleosome and 'beads-on-a-string'

The basic repeat element of chromatin is the nucleosome, interconnected by sections of linker DNA, a far shorter arrangement than pure DNA in solution.

In addition to the core histones, there is the linker histone, H1, which contacts the exit/entry of the DNA strand on the nucleosome. The nucleosome core particle, together with histone H1, is known as a chromatosome. Nucleosomes, with about 20 to 60 base pairs of linker DNA, can form, under non-physiological conditions, an approximately 10 nm 'beads-on-a-string' fibre.

The nucleosomes bind DNA non-specifically, as required by their function in general DNA packaging. There are, however, large DNA sequence preferences that govern nucleosome positioning. This is due primarily to the varying physical properties of different DNA sequences: For instance, adenosine and thymine are more favourably compressed into the inner minor grooves. This means nucleosomes can bind preferentially at one position approximately every 10 base pairs (the helical repeat of DNA)- where the DNA is rotated to maximise the number of A and T bases that will lie in the inner minor groove.

23.19.4 Chromatin and bursts of transcription

Chromatin and its interaction with enzymes has been researched, and a conclusion being made is that it is relevant and an important factor in gene expression. Vincent G., stated that RNA synthesis is related to histone acetylation. The lysine amino acid attached to the end of the histones is positively charged. The acetylation of these tails would make the chromatin ends neutral, allowing for DNA access.

When the chromatin decondenses, the DNA is open to entry of molecular machinery. Fluctuations between open and closed chromatin may contribute to the discontinuity of transcription, or transcriptional bursting. Other factors are probably involved, such as the association and dissociation of transcription factor complexes with chromatin. The phenomenon, as opposed to simple probabilistic models of transcription, can account for the high variability in gene expression occurring between cells in isogenic populations.

23.20 Replication timing

23.20.1 DNA replication

In eukaryotic cells (cells that package their DNA within a nucleus), chromosomes consist of very long linear double-stranded DNA molecules. During the S-phase of each cell cycle (Fig. 23.9), all of the DNA in a cell is duplicated in order to provide one copy to each of the daughter cells after the next cell division. The process of duplicating DNA is called DNA replication, and it takes place by first unwinding the duplex DNA molecule, starting at many locations called DNA replication origins, followed by an unzipping process that unwinds the DNA as it is being copied. However, replication does not start at all the different origins at once. Rather, there is a defined temporal order in which these origins fire. Frequently a few adjacent origins open up to duplicate a segment of a chromosome, followed some time later by another group of origins opening up in an adjacent segment. Replication does not necessarily start at exactly the same origin sites every time, but the segments appear to replicate in the same temporal sequence regardless of exactly where within each segment replication starts. Figure 23.10 shows a cartoon of how this is generally envisioned to occur, while an animation of when different segments replicate in one type of human cell.

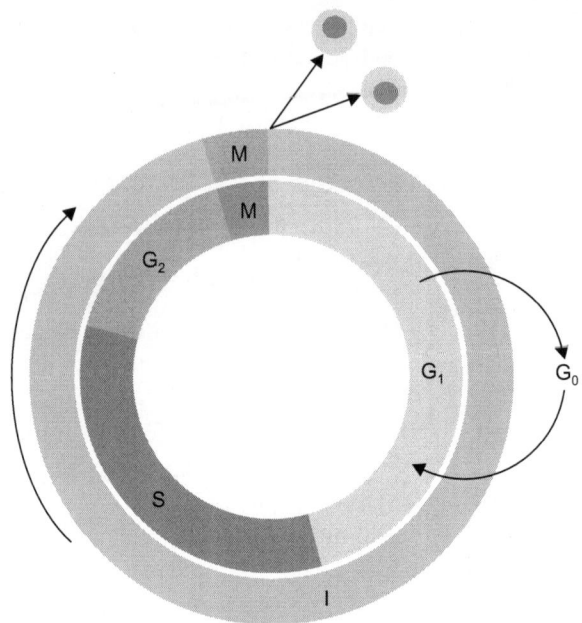

Figure 23.9: Schematic of the cell cycle. outer ring: I = Interphase, M = Mitosis, inner ring: M = Mitosis, G_1 = Gap 1, G_2 = Gap 2, S = Synthesis, not in ring: G_0 = Gap 0/Resting.

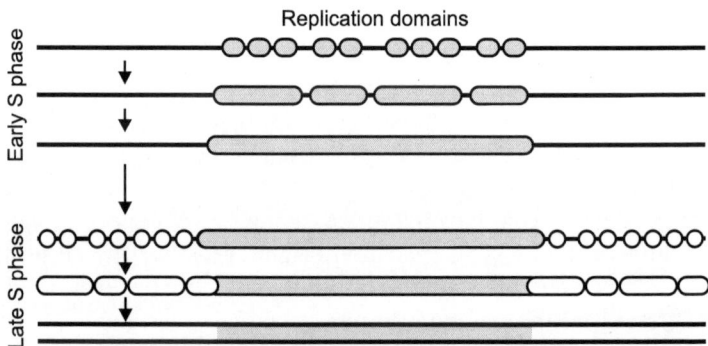

Figure 23.10: Replication proceeds via the nearly synchronous firing of clusters of replication origins that replicate segments of chromosomal DNA ('Replication domains') at defined time periods during S phase.

Replication timing profiles

The temporal order of replication of all the segments in the genome, called its replication-timing programme, can now be easily measured in two different ways. One way simply measures the amount of the different DNA sequences along the length of the chromosome per cell. Sequences that duplicate first, long before cell division, will be more abundant in each cell than the sequences that replicate last just prior to cell division. The other way is to label newly synthesised DNA with chemically tagged nucleotides that become incorporated into the strands as they are synthesised, and then catch cells at different times during the duplication process and purify the DNA synthesised at each of these times using the chemical tag. In either case, we can measure the amount of the different DNA sequences along the length of the chromosome either directly using a machine that reads how much of each sequence is present or indirectly using a process called microarray hybridisation.

Replication timing and chromosome structure

The temporal order of replication of all the segments in the genome, called its replication-timing programme, can now be easily measured in two different ways. One way simply measures the amount of the different DNA sequences along the length of the chromosome per cell. Sequences that duplicate first, long before cell division, will be more abundant in each cell than the sequences that replicate last just prior to cell division. The other way is to label newly synthesised DNA with chemically tagged nucleotides that become incorporated into the strands as they are synthesised, and then catch cells at different times during the duplication process and purify the DNA synthesised at each of these times using the chemical tag. In either case, we can measure the amount of the different DNA sequences along the length of the chromosome either directly using a machine that reads how much of each sequence is present or indirectly using a process called microarray hybridisation.

At present, very little is known about either the mechanisms orchestrating the timing programme or its biological significance. However, it is an intriguing cellular mechanism with links to many poorly understood features of the folding of chromosomes inside the cell nucleus. All eukaryotes have a timing programme, and this programme is similar in related species. This indicates that it is either important itself, or something important influences the programme. It is unlikely that replicating DNA in a specific temporal order is necessary simply for the basic purpose of duplicating a DNA molecule. More than likely, it is related to some other chromosomal property or function. Replication timing is correlated with the expression of genes such that the genetic information being utilised in a cell is generally replicated earlier than the information that is not being used. We also know that the replication-timing programme changes during development, along with changes in the expression of genes. For many decades now, it has been known that replication timing is correlated with the structure of chromosomes. For example, female mammals have two X chromosomes. One of these is genetically active, while the other is inactivated early in development. In 1960, J. H. Taylor showed that the active and inactive X chromosomes replicate in a different pattern, with the active X replicating earlier than the inactive X, whereas all the other pairs of chromosomes replicate in the same temporal pattern. It was also noticed by Mary Lyon that the inactive X took on a condensed structure in the nucleus called the Barr body at the same time during development as the genetic inactivation of the chromosome.

This may not come as too much of a surprise, since the packaging of DNA with proteins and RNA into chromatin takes place immediately after the DNA is synthesised. Therefore, replication timing dictates the time of assembly of chromatin. Less intuitive is the relationship between replication timing and the three-dimensional positioning of chromatin in the nucleus. It is now well-accepted that chromatin is not randomly organised in the cell nucleus, but the positions of each chromosome domain relative to its neighbouring domains is characteristic of different cell types, and after this geography is established in each newly formed cell, the chromosome domains do not move appreciably until the next cell division. In all multi-cellular organisms where it has been measured, early replication takes place in the interior of the nucleus and the chromatin around the periphery is replicated later. Recently developed methods to measure the points where different parts of chromosomes touch each other are almost perfectly aligned to when they replicate. In other words, regions that are replicated early versus late are packaged in such a way as to be spatially segregated in the nucleus, with the intervening DNA containing regions of reduced origin activity. One possibility is that these different compartments within the nucleus, established and maintained without the aid of membranes or physical barriers, set thresholds for the initiation of replication so that the more accessible regions are the first to replicate.

23.21　Gene families and evolution

Gene families refer to two or more genes that come from a common ancestral gene in which the individual members of the gene family may or may not have a similar function. The idea of gene families implicitly invokes a process in which an original gene exists, is duplicated and the resulting gene products evolve. The most common result of gene duplication is that mutation renders one of the products nonfunctional and in the absence of conserving natural selection, one of the members becomes no longer recognisable. Gene families may be clustered or dispersed and may exchange with each other through the mechanisms of gene conversion or unequal crossover. Therefore, understanding the processes of molecular evolution are essential to understanding what gene families are, where they came from and what their function might be. In a sense, duplicate genes allow for more evolutionary potential. At first glance this could be beneficial; if one gene incurred a lethal mutation the other gene simply takes over, there is some protection from mutation based on redundancy. Having two identical genes could result in twice as much product; this may or may not be beneficial in a cell where the integration of thousands of gene products must be coordinated and slight concentration differences can alter biochemical pathways.

23.21.1 Antiquity of gene families

Almost all genes belong to gene families. Evidence from sequence or structural similarity indicates that all or at least large parts of genes came from ancestral genes. Analysis of the human genome has shown that over half of the human genome is comprised of clearly identifiable repeated sequences. Although much of this is owing to self-replicating transposons, i.e. mobile genetic elements, over 5 per cent of the genome has been involved in large segmental duplications in the past 30 million years. If we look further into the past using evidence from protein similarity of three or more genes that occur in close proximity on two different chromosomes, we find over 10,310 gene pairs in 1077 duplicated blocks contain 3522 distinct genes. Because our observations are based on genes that retain similarity, only a small fraction of the ancient duplications can be detected by current means. What is clearly evident is that a very large part of the human genome has come from duplications and that duplication is a very frequent event.

With the evidence showing that gene duplication plays a major role in modern genomes, the question is where did it all begin? How many genes did life start with? Various estimates of the minimal gene set suggest that as few as 250 genes could provide the minimal number of components necessary to sustain independent life. The number of genes in mycoplasms ranges from 500–1500 genes and in bacteria from 1000–4000 genes. Yeast (*S. cervisiae*) has ~6000 genes, worms (*C. elegans*) have ~18,000 genes, the fruit fly (*D. melangaster*) has ~13,000 genes, a plant, *Arabidopsis*, has 26000 genes, and humans have at least 30000 genes. The difference between 250 or 1000 genes and 30,000–40,000 genes is only two orders of magnitude yet the difference in the complexity of life forms seems far greater than could be explained by a simple gene count. Certainly with the increased number of genes comes the opportunity for more complexity in terms of gene interaction. But can synergy alone explain the differences in morphological complexity? Partial explanations invoke more complex differential splicing to account for a higher proportional number of protein products stemming from only 30,000 genes, but one wonders if this observation is merely an artifact of the few numbers of whole genomes available to us.

23.21.2 Origins

Gene duplications can come from a variety of sources including whole or partial genome duplication. Polyploidy results from a failure of chromosome segregation during the cell division of gametes. The most distinguishing feature of polyploidy is that it effects all of the genes simultaneously so that the relative proportion of genes within cells remains the same. Among plants and invertebrates, polyploidy is quite common and in many species it has little effect on phenotype. Ohno has argued that whole genome duplications are the most important events in evolution yet others have suggested that polyploidy has no effect on phenotype. More recent discussions: acknowledge the potential that polypoidy brings to gene family evolution but also appreciate the role of complexity of gene interactions in determining the impact of whole genome duplications. In vertebrates, polyploidy is quite rare. Most of the 188 examples of genome duplication have been found in amphibians, reptiles and some fish (salmon). In these instances, polyploid species have undergone dramatic changes to reestablish diploidy through chromosome loss, mutation and rearrangement. Tetraploid genomes have no trouble going through cell division as long as chromosomes remain very similar. But as mutations arise and duplicated chromosomes begin to differ, cell division can no longer insure equal division of genetic material to germ cells and severe imbalances can occur during chromosomal segregation. The initial transition phase from tetraploid to diploid results in huge losses of gametes and developing young. In salmon it is estimated that approx 50 million years after a polyploid change, only 53 per cent of duplicate genes remain. The result of duplication by whole or partial genomes can result in large changes in gene number but there are major difficulties in cell replication that must be overcome.

The rapid increase in genome size through polyploid events has been used to explain the increased size of mammalian genomes. Ohno suggested that two rounds of genome duplication occurred early in vertebrate history. This may explain the Cambrian explosion in which vertebrates appeared in paleontological records quite rapidly. Evidence for two rounds of genome duplication comes from vertebrates having four times the number of developmental regulator genes (Hox, Cdx, MyoD, 60A, Notch, elav, btd/SP.) as *Drosophila*. While this concept has become very popular in the literature, recent studies examining expected phylogenetic relationships among genes have called into question whether the number of genes was a result of two genome-wide duplications or simply a result of ongoing, frequent genome segment duplications. While the primary support for quadruplicated genomes comes primarily from chromosomes 2, 7, 12 and 17, which contain the Hox gene clusters, a more extensive examination of the number of homologous genes within human as compared to genes within *Drosophila* was unable to resolve the question of whether two whole genome duplications gave rise to modern vertebrate genomes.

23.21.3 Mechanisms

Duplication of large blocks of DNA cannot be explained by chromosomal segregation errors. Mechanisms of large segmental duplication are varied and the role of transposable elements in gene duplication is often cited as a primary cause. Transposable elements and in particular retrotransposable elements are highly repetitive dispersed sequences that can replicate independent of nuclear division. In human they comprise over 45% of the genome. While there are a number of instances where retroposons have been found at the junctions of duplicated segments, there are also a number of instances where they have not. To understand why gene segment duplications appear common, it is perhaps important to look at the DNA molecule itself. DNA is composed of duplex strands held together by hydrogen bonds whose strength varies. In a fluid environment, various local salt concentrations, temperatures, physical torsion forces and local nucleotide compositions (e.g., levels of G + C, simple repeats) can result in temporary separation of strands of duplex DNA. If similar sequences are found in the same physical location, unstable heterologous duplexes can form. Heterologous pairing or single-strand conditions are prone to stress and breakage. These situations are repaired correctly in the vast majority of cases but occasionally mistakes are made that result in new gene neighbours. The possibility of error is particularly high during cell division when DNA is being replicated and when similar sequences are in close proximity. The potential impact of highly repeated transposable elements as a destabilising factor and a potential focal point for rearrangement becomes clear in our genome. It is, therefore, somewhat surprising to find that many of the duplications are not flanked by repeat elements. What is clear is that duplication involves local chromosome instability that results in breakage and aberrant repair of the ends. Factors that promote segmental duplication include close proximity and high sequence similarity. It then follows that duplications resulting in adjacent genes would be more susceptible to further changes than duplications resulting in dispersed genes. Furthermore, adjacent duplications create far fewer chromosomal segregation problems during cell division and therefore, should be found more often in the genome. What is observed is that large segmental duplications involving multiple genes are dispersed throughout the genome whereas duplications involving single gene segments are both dispersed and in close proximity.

There are many instances of clustered gene families (globins, Hox, Ig, Tcr, Mhc and rRNA). Among tandemly duplicated gene segments there is the possibility of extensive gene conversion and unequal crossing over (Fig. 23.11).

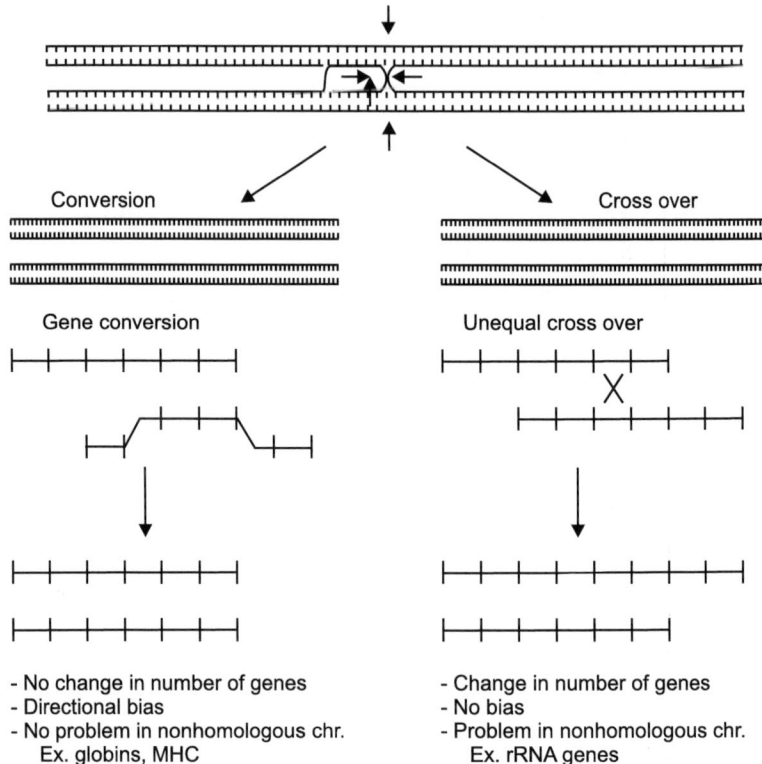

Figure 23.11: Gene conversion and unequal crossing over mechanisms of communication among gene family members. The arrows indicate possible break points that would result in either conversion or cross over results.

The later is the predominant mode of change. Unequal crossing over between dispersed genes results in extreme difficulties in chromosomal segregation in cell division but gene conversion does not. Gene conversion replaces the sequence of one family member with the sequences of another close (>90% similarity) member but it does not effect the total number of genes. Gene conversion requires DNA strand breakage followed by strand migration to a similar gene and the formation of a heteroduplex. DNA repair mechanisms then repair differences in the heteroduplex often using one strand corresponding to the unaffected homologous chromosome as a template. The heteroduplex then resolves and may go back to its original location carrying with it DNA changes. Heteroduplex formation is often temporary and most often occurs between alleles of the same gene though occasionally it may affect paralogous genes in which more than 100 bases are greater than 95% similar. Depending on the resolution of the heteroduplex and biases in mismatch repair, adjacent base differences may both reflect one or the other parental strand or they may reflect a combination of parental strands. The end result is that the total genetic variation is reduced but a particular gene may increase its number of alleles. Genetic variation at one gene can increase over a single conversion event, but over multiple conversion events variation is reduced. The resolution of hetero-duplexes formed from the invasion of a DNA strand from one gene segment into the duplex of a similar, adjacent duplicate can also result in unequal crossing over. Unequal crossing over changes the number of genes. For example, in a tandem arrangement, unequal crossing over results in one chromosome with one duplicate and the other chromosome with three duplicates where the front and the back parts of the single duplicate and the middle duplicate of the triplicated segment reflect different origins. Figure 23.12 shows three successive unequal crossing-over events and shows the expected phylogenetic relationships of the front (5′) and back (3′) parts of the gene. It is evident from the final trees for the 5′ and 3′ parts of a tandemly arrayed gene segment that it is possible to describe some of the major, more recent evolutionary events that have occurred. Because information is lost due to the fixation of one of the cross over products in each population, it may never be possible to obtain a complete historical

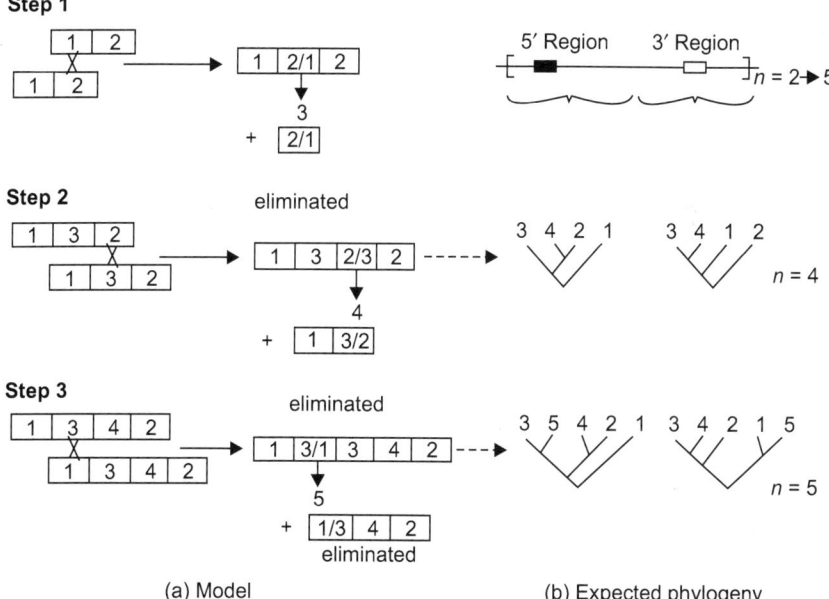

(a) Model   (b) Expected phylogeny

Figure 23.12: Unequal crossing over between tandemly arrayed gene family members. This model: (a) assumes a break-point near the middle of the duplicated segments. The expected phylogeny (b) represent sequence relationships between the 5′ and 3′ regions of the duplicated gene segments.

picture. But we can see in examples from the literature that unequal crossing over is the major factor in clustered gene families.

23.21.4 Variation

Genetic variation increases when the number of duplicates increases but it is decreased when the number of duplicates decreases. It is important to remember several tenets of unequal crossing over.

First, the ultimate fate of duplicates undergoing multiple unequal crossing overs is to return to a single copy unless selection maintains multiple copies.

Second, while the overall variation may increase over a single event, the result over multiple expansions and deletions is homogenisation of duplicates (example rRNA genes). Third, unequal crossing over between dispersed gene segments often results in fatal problems in cell division.

Lastly, unequal crossing over is the predominant mechanism that increases or decreases the number of gene family members in clusters. It appears that the factors that promote duplication include proximity, high similarity, larger numbers of existing duplicates and internal sequences that are prone to breakage.

Given these factors, it is perhaps surprising that we do not see more evidence of repetitive elements playing a larger role in gene duplication. At the same time it becomes easy to see the complex evolution and interactions among both dispersed and clustered gene families.

23.21.5 Genes and domains

Duplication can involve very large stretches of DNA, whole genes or even parts of genes. Of 1077 duplication blocks containing three or more genes in the human genome, 159 contained 3 genes, 137 contained 4 genes and 781 contained five or more genes. At the same time we often see clusters of gene family members. This indicates that duplications often involve one gene or even parts of genes. Clearly the mechanisms of duplication outlined earlier play a major role at all levels of gene family evolution. However, it is important to remember that the events that are most evident are those that are fairly recent or those that involve conserved genes. Sequence similarity for older duplications of noncoding DNA rapidly fades. It is our focus on function that draws us to study genes. As mentioned, several genes within larger segments can be duplicated but perhaps just as interesting, parts of genes (introns and groups of introns) can be duplicated. This is particularly interesting because genes are composed of functional domains. Remarkably, there may be fewer than 1000 classifications in existence. Domains can be mixed, matched, duplicated and modified to provide novel functions within genes as well as between genes. Only 94 of the 1278 protein families in our genome appear to be specific to vertebrates. That may be an overestimate resulting from our inability to recognise similarity. It appears that the 30000-plus genes in the human genome are not novel but simply products of duplications and mixing and matching of existing genes and domains to create new genes and new functions.

23.21.6 Species evolution and gene evolution

The study of gene evolution is incomplete without the study of species evolution. Gene evolution and species evolution is not the same but knowledge of one greatly benefits our knowledge of the other. Modern species are the result of a dense network; of transient species that occasionally give rise to other species. Using paleontological records as well as morphological, physiological and developmental studies of extant and extinct life forms, we are able to trace some of the origins of modern species. But numerous gaps in our understanding remain. Gene evolution can occur within a species but when a speciation event occurs, gene evolution within each new lineage is independent of gene evolution in other lineages. For example, gene evolution within human is independent of gene evolution within chimpanzees.

To illustrate this point, Fig. 23.13a shows a gene duplication in a common ancestor of species A and B followed by a speciation event and separate A and B lineages. Figure 23.13a is the corresponding gene phylogenetic tree. Note that the timing of speciation events can be used to time major events in gene evolution. In Fig. 23.13a, gene duplication occurred before the speciation event. In Fig. 23.13b, a more complex gene evolution is shown. Within each species one of the genes has been eliminated and the other has been duplicated so that each species has two genes, which appear to have arisen after the speciation event. While many other scenarios can occur, these two illustrations show how even in relatively simple cases, one must be cautious when interpreting gene trees. A number of studies have used these methods to identify new gene family members. Slightom was one of the first to use combined gene and species studies to examine genetic mechanisms of change. What is clear is that the use of both gene trees and species trees can be a very powerful method of studying species evolution and gene evolution.

23.22   Repeated sequence (DNA)

Repeated sequences (aka repetitive elements or repeats) are patterns of nucleic acids (DNA or RNA) that occur in multiple copies throughout the genome. The functions and descriptions of these sequences are currently being characterised by scientists.

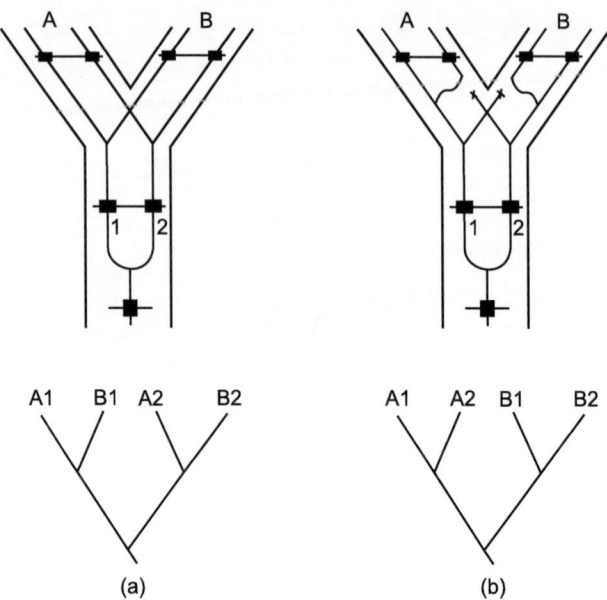

Fig. 23.13: Species evolution (in bold outline) and gene evolution (light lines). (a) Shows a gene duplication in a common ancestor of species A and B followed by a speciation event and separate A and B lineages and (b) A more complex gene evolution is shown. Within each species one of the genes has been eliminated and the other has been duplicated so that each species has two genes, which appear to have arisen after the speciation event.

23.22.1 Types of repeats

There are 3 major categories of repeated sequence or repeats:

- Terminal repeats.
- Tandem repeats: copies which lie adjacent to each other, either directly or inverted.
    - › Satellite DNA typically found in centromeres and heterochromatin.
    - › Minisatellite repeat units from about 10 to 60 base pairs, found in many places in the genome, including the centromeres.
    - › Microsatellite repeat units of less than 10 base pairs, this includes telomeres, which typically have 6 to 8 base pair repeat units.
- Interspersed repeats (aka interspersed nuclear elements).
    - › Transposable elements (aka. transposons, or retroelements).
    - › Short Interspersed Nuclear Elements (SINEs).
    - › Long Interspersed Nuclear Elements (LINEs).

In primates, the majority of LINEs are LINE-1 and the majority of SINEs are Alu's. In prokaryotes, CRISPR are arrays of alternating repeats and spacers.

23.2.2   Other types of repeats

- Direct repeats:
  › Global direct repeat.
  › Local direct simple repeats.
  › Local direct repeats.
  › Local direct repeats with spacer.
- Inverted repeats:
  › Global inverted repeat.
  › Local inverted repeat.
  › Inverted repeat with spacer.
  › Palindromic repeat.
- Mirror and everted repeats.

Section IV

Role of nanotechnology in derivatives of molecular biotechnology

Nanotechnology in cancer diagnosis and treatment

24.1   Introduction

Definitions of nanotechnology vary widely. Some scientists restrict the definition to work with molecules and devices between 1 and 100 nm, while others widen these parameters to 1–1000 nm. However, nanotechnology is best defined by the capacity to artificially construct and manipulate structures at nanoscale and nanoscale's novel properties. Drug loaded nanoparticles are widely utilised in the treatment of a number of diseases, such as metabolic disorders, autoimmune diseases, inflammatory disorders, neurodegenerative diseases and cancer. For instance, nanomedicines have been extremely useful in improving the efficacy of small molecule drug delivery across the blood-brain barrier for the treatment of Central Nervous System (CNS) diseases. In addition, nanoparticles serve as artificial oxygen carriers that can act as a substitute for blood, saving the lives of those in dire need of transfusion. Although liposome-encapsulated formulations of Doxorubicin were being widely administered as early as the 1990's, nanotherapeutics is still viewed as a new and emerging field. This chapter focuses on the progress made using nanoparticles in cancer prevention, diagnosis and treatment. This is certainly an area of rapid progression, with current nanotherapeutics for cancer encompassing a vast array of nanomaterials and nanodevices.

However, much investment, research and development into nanotechnology diagnostics, therapies, devices, biosensors and microfluidics continues to provide advances in the prevention, diagnosis and treatment of cancer. Many scientists believe that nanoparticles are the future of diagnosis and drug delivery with the potential to overcome many of the obstacles that cancer presents.

24.2   Obstacles in cancer diagnosis and treatment

24.2.1   Late stage diagnosis

Late detection and diagnosis of cancer remains one of the fundamental causes of low survival rates, so developing a test that detects clinically apparent cancer before symptoms appear is obviously an important goal. The traditional biomedical imaging tools of magnetic resonance imaging, ultrasound and positron emission tomography have several limitations in the diagnosis of cancer, including an inadequate imaging period, a risk of renal toxicity and an inability to detect tumour cells smaller than 1 cm. Improvements in PET, CT and MRI, through the use of small molecule imaging agents, such as 2-deoxy-2-(18F) fluoro-D-glucose [FDG], iodinated small molecules and chelated gadolinium respectively, are routinely used in the diagnosis of cancer. However, poor stability, rapid clearance and low signal intensity have limited the use of these techniques and prompted more research into the use of nanoparticles as a diagnostic tool.

24.2.2 Challenges in targeting, transport and delivery of treatment

Chemotherapy's perennial problem has always been that, due to challenges presented by its targeting, transport and delivery, a pharmacologically active concentration in tumour cells is often only achieved at the expense of what Couvrer terms 'massive contamination of the rest of the body'. This toxicity can result in the use of suboptimal and/or intermittent dosing, to allow the body to rest, or in some cases to forgo chemotherapy altogether.

Many traditional chemotherapeutics have poor stability and aqueous solubility. Due to this limitation, many drugs, despite significant biological activity, are disregarded at early stages of drug screening in the laboratory. In addition, distribution of some drugs is too general, with only a small fraction of drugs reaching the cancer site; injected agents are often cleared by the monocytes and macrophages of the reticuloendothelial system (RES). To be successful, a therapeutically sufficient quantity of the drug, still in a viable state, must survive clearing and be delivered to different regions of tumours, via., blood vessels, cross the vessel wall and then finally penetrate through the interstitial space to reach the target, where unpredictable blood flow and often abnormal vasculature in tumours, particularly in necrotic and semi-necrotic regions, can make accurate delivery even more difficult.

Other than conventional chemotherapeutic drugs, biological molecules, such as antibodies and nucleic acids, are being widely explored for the treatment of different diseases, including cancer. Nucleic acid drugs, such as aptamers, anti-sense DNA/RNA and small-interfering RNA, have shown great promise in the treatment of cancer. However, these drugs are greatly limited by serum nucleases, opsonisation and clearance by macrophages and clearance by the renal system. Some of the nucleic acid therapeutics, such as stable nucleic-acid–lipid particle (SNALP), have used nanocarriers to effectively overcome the above mentioned barriers and are being used in the treatment of cancer.

24.2.3 Chemoresistance

Chemoresistance, a major cause of cancer treatment failure, is linked to cancer stem cells (CSC). CSCs possess unique properties, such as quiescence, mesenchymal morphology, increased DNA repair ability, overexpression of antiapoptotic proteins, drug efflux transporters and detoxifying enzymes. These properties, together with the favourable tumour microenvironment and hypoxic stability, mean that they often escape elimination by current radio and chemotherapies. Having survived through chemotherapy, they can give rise to metastases and recurrent tumours which then increase in malignancy and resistance.

Chemoresistance can be divided into two types: Intrinsic and acquired. Intrinsic chemoresistance is displayed by tumour cells whose genetic and phenotypic characteristics make them ideally suited to withstand cytotoxic agents. Acquired chemoresistance can occur after prolonged exposure to chemotherapeutic agents, which disrupt only one of the many biochemical pathways involved in their pathogenesis. Unfortunately this approach often activates and strengthens the alternative pathways, resulting in chemo resistant mutations in the tumour cells and tumour relapse. Multidrug resistance, or MDR, can also occur through a process of cross-resistance in which cancer cells mutate and acquire resistance to multiple structurally-related drugs and also to mechanistically different drugs, either, via., the over-expression of multidrug transporters or through altered apoptosis, resulting in decreased intracellular drug retention and altered tumour response.

24.2.4 Patient—compliance and individuality

It may seem harsh to list the patient as an obstacle but, through no fault of their own, this is often the case. Genetic variation across individuals affects a drug's pharmacokinetics and pharmacodynamics and a breakthrough with one patient cannot always be replicated in another. Also, each patient has different levels of tolerance to the discomfort and effects of chemotherapy and in many cases a pre-existing condition or illness may complicate their cancer treatment or lead them to refuse it. Patients with comorbid illnesses and elderly patients are those most likely to forgo or discontinue chemotherapy.

24.3   Current nanotherapeutics: Overcoming the obstacles

24.3.1 More accurate detection and diagnosis

Promising results have emerged from combining nanoparticle-based optical contrast agents with existing optical imaging technologies. Their 'programmable' surface properties and potential for passive or active targeting make nanoparticles ideal for diagnostic imaging. The ability of nanoshells, constructed with a silica core and gold shell, to absorb specific wavelengths of light, has great potential for cancer imaging and therapeutic applications.

24.3.2 Quantum dots

Semiconductor quantum dots are luminescent nanocrystals with great potential in both biological and biomedical applications. Their photostability, fluorescence intensity, small size (2–10 nm) and tunable surfaces make them ideal for optical imaging and detecting hundreds of cancer biomarkers in blood assays or tissue biopsies at pg/mL concentrations. The most commonly used agents in the quantum dots (Fig. 24.1) are selenides or sulphides of cadmium and zinc.

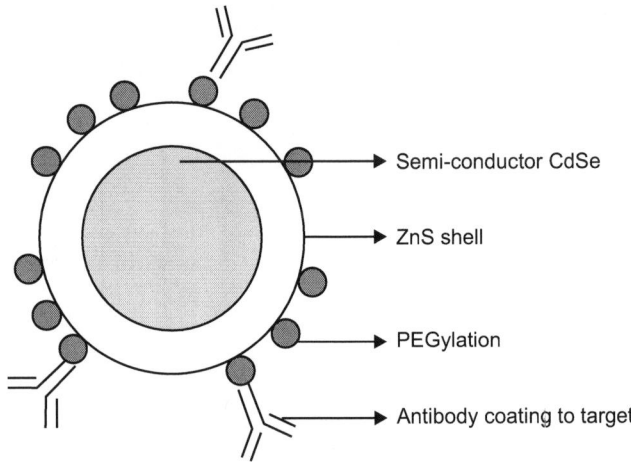

Semi-conductor CdSe

ZnS shell

PEGylation

Antibody coating to target

Figure 24.1: Quantum dots.

The wavelength of light emitted by the quantum dots depends on their size. The light emitted is much more intense and stable than their other fluorescent counterparts and hence very useful in optical imaging. Cadmium selenide (CdSe), cadmium telluride (CdTe), indium phosphide (InP) and indium arsenide (InAs) are the most common quantum dot formulations used in biological applications. The inorganic core is covered by an inorganic shell, which imparts greater photostability and increases the fluorescence properties of the core. The surface of the shell is coated with another layer that enhances solubility and stability of quantum dots in the blood. Often times, the surface coating is PEG as it has low toxicity and is biodegradable. A major limitation of quantum dots in imaging is a process called 'Blinking'. This is due to fluctuation of the quantum dots between the light emitting and non-emitting states. This limits the amount of signal obtained at a specific time.

24.3.3 Magnetic resonance imaging

Recently, the development of nanoparticle systems to improve MRI for cancer imaging and diagnosis has made significant progress. Magnetic nanoparticles usually consist of an inorganic nanoparticle core and a surface coating that provides stability in aqueous dispersions. This surface coating is manipulated to facilitate targeting, real-time monitoring or both. Their success, particularly as contrast agents for MRI, is largely due to their enhanced proton relaxation and deep-tissue imaging capabilities, non-invasiveness and low toxicity.

Supermagnetic iron oxide (SPIO) nanoparticles are now widely used as bowel contrast agents and have been used for some time in spleen/liver imaging. SPIO nanoparticles are readily taken up by macrophages present in the liver parenchyma (Kupffer cells) and as liver tumours are usually devoid of macrophages, the macrophage-specific uptake of SPIOs increases the contrast between healthy and diseased tissue, allowing liver tumours or micro-metastases as small as 2–3 nm to be detected. SPIO nanoparticles are biodegradable as the iron molecules released into the plasma upon degradation can bind haemoglobin. To avoid clearance of the SPIO nanoparticles, they are often coated with PEG, which enhances the circulation time during the imaging and treatment of prostate cancer. They are conjugated with an aptamer that binds with high specificity and affinity to a cell surface ligand on the prostate tumour cells. The aptamer binding to the cell surface antigen cause a localised increased accumulation of the SPIO that enables imaging. In addition, conjugation of doxorubicin to the SPIO nanoparticles allows the targeted delivery of the chemotherapeutic drug with minimal side effects.

24.3.4   Molecular diagnostics

AuNPs (gold nanoparticles) have brought great benefits in this area, with increased sensitivity and specificity, multiplexing capability and short turn around times. Aptamer-conjugated NPs can also be used for the collection and detection of multiple cancer cells. Gold nanoparticles scatter light intensely. Based on the size and shape of the gold nanoparticles, the scattering properties of the gold nanoparticles are also changed. The light scattered by the gold nanoparticles have greater photostability than the other imaging agents commonly used. Gold nanorods exhibit a phenomenon called the Surface Enhanced Raman Scattering (SERS), which is also used in cancer diagnosis. In addition, gold nanoshells and gold nanorods have been used to induce photothermal therapy. This is an example for a 'theragnostic' agent, as the gold nanorods not only assist in diagnosis of the cancer, but also help in ablation.

24.3.5   Improving targeting, transport and delivery

Nanoparticles are increasingly utilised because of the multiple benefits that they offer. Nanodelivery systems can make the use of chemotherapy drugs more safe and efficient by improving delivery, cell uptake and targeting and have been shown to improve their pharmacokinetic profiles and enhance their targeting at the required site. This success relies on two main factors: (i) the EPR and (ii) the potential ability of nanodrug delivery systems to overcome the shortcomings of many anticancer drugs.

The EPR, or enhanced permeation retention effect, exists because of two properties of tumours. Firstly, tumour tissues have increased vasculature which allows the entry of macro-molecules and colloidal particles of diameter up to 600 nm. Secondly, the lymphatic system is not effective in clearing the interstitial fluid from the tumour tissues. Normal tissues other than the spleen, liver and kidney are impermeable to molecules that are larger than 2nm. Hence, nanoparticles can selectively target tumour tissues reducing toxic side effects. Together, the enhanced permeation and retention properties of the tumour over the normal tissues cause the nanoparticle to have prolonged contact with the tumour cells. In addition, nanocarriers also release the drug slowly, ultimately resulting in reduced drug distribution and toxicity to normal tissues.

Once the nanoparticles reach the target tissue, cell surface receptors interact with ligand-coated nanoparticles leading to their uptake by endocytosis. Cellular uptake of uncoated nanoparticles is governed by their differences in size, shape and charge. It is suggested that positively charged nanoparticles are taken up more readily due to electrostatic attraction. Interaction with specific serum proteins, results in the formation of a corona, promoting cell entry. Recent studies indicate that non-spherical molecules, such as rod-shaped structures, are internalised better than spherical structures. Uptake of larger nanoparticles disrupts the membrane surface, thereby inducing cell death. Non-specific uptake of the nanoparticles by the lung epithelial cells and red blood cells could cause toxic side effects.

Nanosized drug delivery systems can potentially overcome the shortcomings of many anticancer drugs, such as low aqueous solubility and stability and high nonspecific toxicity. For example, paclitaxel nanoparticles stabilised with pluronic F68 are stable for years, while the same drug in dissolved form degrades completely in less than 48 hr and magnetic nanoparticles (MNPs) are increasingly used because their targeting ability reduces systemic distribution of cytotoxic compounds *in vivo* and enhances uptake at the target site, resulting in effective treatment at lower doses.

24.3.6   Nanodrug delivery systems

Nanodrug delivery can either use passive targeting mechanisms, such as the EPR effect, or active targeting mechanisms, using ligands directed against differentially overexpressed cell surface markers surface on tumour cells. Drug encapsulation within nanoparticles can also enhance the bioavailability of drugs administered, via., routes other than intravenous; both insoluble and soluble drugs can be incorporated within nanoparticulate sols, extending their stability as they travel through the blood, which in turn improves their overall pharmaco-kinetic half-life.

By 'pre-programming' the degradation of nanoparticles in the body, prolonged drug release can be achieved, eliminating the need for repetitive dosages and enabling more sustained and consistent drug concentrations in the target area. Brannon-Peppas and Blanchette have compared the uptake of nanoparticles with more hydrophobic surfaces with those of more hydrophilic surfaces. They concluded that a nanoparticle designed to be 100 nm or less in diameter with a hydrophilic surface will have a longer circulation time and hence a greater ability to target the required site due to reduced clearance by macrophages. Nanodrug delivery systems can carry one or a combination of therapeutics, including cytotoxic agents, chemo sensitisers, small interference RNA (siRNA) and antiangiogenic agents. The most commonly used platforms are described below.

Liposomes (Fig. 24.2) have been in use for the past several decades and are established as drug and imaging agent carriers with proven clinical efficacy. They are artificial phospholipid vesicles 50 nm to $= 1\ \mu$m in size, either unilamellar or multilamellar (with one lipid layer or several, arranged in onion-like layers), with one or more aqueous compartments.

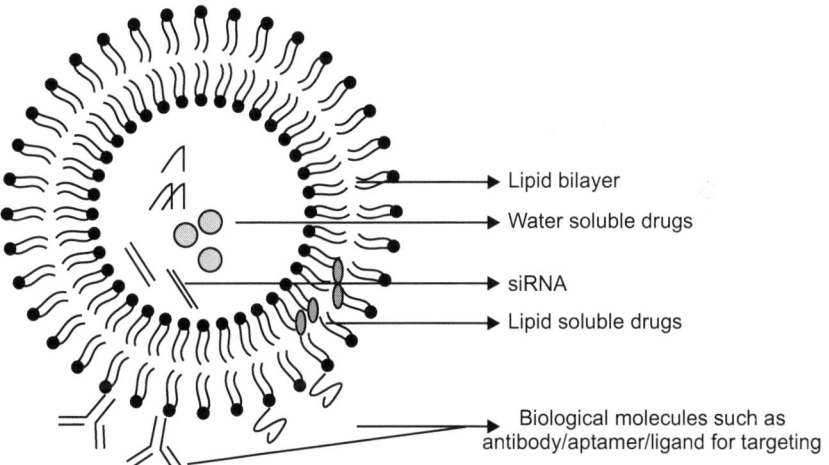

Lipid bilayer

Water soluble drugs

siRNA

Lipid soluble drugs

Biological molecules such as antibody/aptamer/ligand for targeting

Figure 24.2: Liposome.

The 'cargo' can be held in the aqueous compartment(s) or lipid layer. Liposomal nanocarriers provide protection from degradation. Optimisation of the pharmacokinetics of the encapsulated drug can improve drug accumulation in the tumour and reduce the adverse effects of bolus administration. Polymeric micelles consist of a hydrophobic core and a hydrophilic shell and are useful drug carriers, due to their tunable size and surface functions, high monodispersity and excellent stability. They have the ability to form hydrogels and are used for drug encapsulation or drug conjugation. Under the right conditions, pluronics, the most well-known thermosensitive polymers, form a hydrogel at body temperature but are water soluble at 2–4°C. This allows them to be injected as a liquid but they form a hydrogel *in situ*, resulting in prolonged drug release of the encapsulated drug.

Dendrimers are globular macro-molecular compounds consisting of an inner core, which can be manipulated to alter its shape and size, surrounded by a series of branches with surface functional groups. They can carry a multiple payload of active targeting molecule, diagnostic agent and therapeutic drug and those with a hydrophobic core and hydrophilic surface groups can form micelles, which can then be designed for site-specific release of their payload, via., pH and enzyme dependent mechanisms.

Inorganic nanoparticles, such as gold nanoparticles, can be used as a cargo for drug delivery. Gold has a number of appealing surface properties, such as light scattering, which makes them attractive inorganic biomaterials for drug delivery when combined with nanoparticles. Due to their ease of synthesis, biocompatibility and affluent functionalisation, many drugs can be conjugated to the surface of gold through hydrophobic interactions. Gold–thiol conjugates are the most common due to their accurate and predictive functionalisation. Various antibiotics, anticancer agents and oligonucleotides are also conjugated with gold nanoparticles to yield more viable drug delivery agents. Other inorganic nanoparticles that are frequently used in drug delivery and diagnostics include silica and iron oxide, which forms the core constituent of many inorganic nanoparticles.

Porous silica based nanoparticles are highly suitable for carrying hydrophobic drugs. These nanoparticles have a high surface-to-volume ratio and consists of large pores. The drug can be loaded on the nanoparticles by physical adsorption and covalent linkage. Nanostructured mesoporous silicon (PSi), fabricated by electrochemical etching, have nanometer range pores that facilitate high drug loading capabilities, irrespective of different surface chemistries.

24.3.7 Nanoparticles as therapeutic agents

Nanoparticles can be used as a therapeutic agent themselves. Their ability to alter the substrate molecule, through a process called 'intercrossing' upon excitation by light, is used to treat cancer cells in the photodynamic therapy. While in the photothermal therapy, the property of small inorganic molecules to generate heat upon excitation is taken advantage of in the inducing apoptosis or necrosis of cells.

Photothermal therapy

Photothermal therapy (PTT) uses sensitisers that can absorb light in the near-infrared region and convert it to thermal energy, causing heat in the vicinity. The sensitiser used in PTT is usually inorganic molecules, such as gold or carbon nanoparticles. Thermal ablation therapy has been used in the treatment of cancer for many decades, but the damage to nearby tissues has limited the use of this technique in the treatment of cancer. However, with the advent of photodynamic therapy (PDT), targeted destruction of the tumour cell has become possible. PEGylation and active targeting of Au nanotubes have been used in the treatment of many cancers.

Photodynamic therapy

Photodynamic therapy (PDT) uses photosensitisers in the treatment of cancer or other disorders. Photosensitisers are molecules that can be excited by light, which then alters molecules in the vicinity, causing the release of singlet oxygen species reactive oxygen species (ROS). ROS are capable of causing oxidative stress to the surrounding cells, causing apoptosis or necrosis. Photosensitisers can be excited using lasers over a wide range of visible wavelengths. Because of the limited penetrability of visible light, photosensitisers can be used to treat only superficial tumors, such as skin, lung, esophagus, prostate, head and neck, colon and rectum to mention a few.

Because the half-life of the reactive oxygen species is only a few milliseconds, this therapy can be used to cause targeted cell death in regions where the photosensitiser has accumulated. Photosensitisers can be coated with polyethylene glycol to prevent renal clearance and to enhance the circulation time in the blood. Further antibody conjugation to the surface can target the photosensitiser to the cancer cells that overexpress the antigen on the surface. Figure 24.3 shows combinational phototherapy.

24.3.8 Overcoming resistance

Nanovehicles carrying therapeutic drug combinations that not only target the tumour cells selectively, but also overcome the mechanisms of drug resistance are the focus of intensive research. This method has been proved especially effective in circumventing multidrug resistance (MDR) in multiple cancer models. MDR was reversed in *in vitro* and *in vivo* cancer models through the co-delivery of combinations of chemo sensitising agents and chemotherapeutic agents.

24.3.9 Improving therapy

Nanoshells are nanoparticle beads with a thin gold outer shell and a central silica core. By manipulating the thickness of the shell and core, the beads can

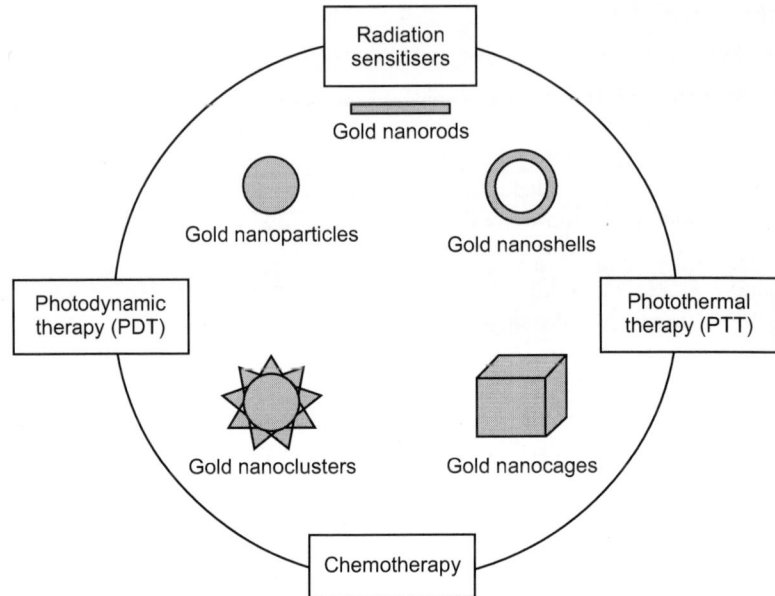

Figure 24.3: Combinational phototherapy.

be tuned to absorb and scatter specific wavelengths of light across the visible and near-infrared (NIR) spectrum, which is very useful in enhancing imaging properties. Arguably, however, this ability to absorb light is most usefully exploited in thermal ablation therapy. For maximum efficacy, nanoshells with a silica core diameter of ~120 nm and a 10-nm gold shell are used in this therapy as they strongly absorb NIR light (~800 nm) and can then create intense heat that is fatal to cells. As tissue chromophores do not absorb much energy in the NIR range, NIR light can penetrate several centimeters of human tissue without causing harm.

24.3.10 Improving cancer prevention

The complete prevention of cancer occurrence, claims Smith and others as an unachievable goal; cancer prevention describes 'slowing the process of carcinogenesis' and inhibiting its reoccurrence. Inefficient systemic delivery and bioavailability of chemopreventive agents has so far limited their applicability to human medicine. However, Smith and others have experimented with encapsulating a chemopreventive agent, epigallocatechin-3-gallate (EGCG), in polylactic acid and polyethylene glycol (PEG) nanoparticles. Nano-EGCG had a significantly longer half-life and had more than a 10-fold dose advantage over nonencapsulated EGCG in cell growth inhibition, proapoptotic and angiogenic inhibitory effects. Curcumin derived from turmeric, when conjugated with polymeric amphiphile, mPEG-PA or PEG, has been shown to have more significant antiproliferative effects than the free curcumin. Another nanoparticle based formulation, called solid lipid nanoparticles (SLN), is also being used as newer therapeutic modality to address the area of chemoprevention. The advantage is that they act like colloidal carriers which remain as solids at room and body temperature and so can be efficiently used as alternatives to lipososmes and other polymeric nanoparticles. A multitude of approaches utilising nanoparticles to combat these existing deficits in the chemopreventive strategies will re-captivate the 'silver bullet' for chemoprevention in the near future.

24.3.11 Improving compliance

Nanotherapeutics can be less invasive than conventional diagnosis and treatment methods. This leads to shorter recovery times and a decreased risk of infection and these advantages in turn should lead to a reduction in cost and improved life expectancy and quality.

24.4 Future: Potential risks, rewards and research

24.4.1 New risks: The voice of caution

With its obvious potential for breakthroughs in so many fields, it is easy to view nanotechnology as an exclusively positive concept. However, it is not without risk and nanomaterials may present greater risks than their larger counterparts, as their greater relative surface and unique quantum effects mean they have a tendency to be more active and reactive. Their potential to cause harm is harder to predict, as it is determined using factors such as surface area, rather than molecular structure, which is used to risk assess most other chemical hazards and there are no proven toxicity screening methods to evaluate them. The scarcity of information about how nanomaterials may impact safety, health and the environment, along with the growing number and diversity of nanotechnologies and their associated engineered properties, has raised serious concerns. If nanomaterials escape the laboratory or manufacturing site, their degradation and interaction with substances in the environment would be unpredictable and potentially hazardous.

When assessing the risk to patients, it is important to bear in mind that preclinical trials of nanodrugs may be less indicative of human risks than trials of standard medicines and that nanomaterials can utilise unique mechanisms and routes of exposure, potentially by-passing the blood-brain barrier. If inhaled, they may aggregate in the alveoli, where their increased surface area places a burden on mucociliary and macrophage clearance.

Like any other emerging area of interest in human health, nanotechnology also has its own demerits. A word of caution is that this research is still in its infancy to determine the unforeseen side effects pertaining to nanoparticle related therapies. Although our understanding on the concepts regarding nanotherapeutics has come a long way, the exact nature of nanoparticulate drug interactions has not been tested vigorously. Studies in animals suggest alarming facts affecting the brain function. With the limited current literature in humans, it is almost impossible to judge their safety over efficacy.

Hence, until a stringent risk assessment strategy is employed, nanotherapeutics should not be viewed exclusively as a positive concept. It is important that, if nanotechnology is to move forward safely and sustainably, a thorough assessment of the biocompatibility and toxicity of nanoparticles is undertaken, with potential toxicities identified and their underlying mechanisms understood. Research into the avoidance of health risks associated with nanotechnology may potentially be used to guide therapy and *vice versa*.

24.5 New rewards: A bright future

24.5.1 Multifunction nanovehicles

Advances in MRI contrasting agents promise a next generation of agents consisting of a core and coating conjugated to tumour-specific moieties for improved efficacy and tumour targeting. Perche and Torchillin have suggested that a possible direction for research may be the coupling of ligands of different natures (antibodies, proteins, peptides and chemokines, hormone analogs) to target at least two tumor cell populations, providing more sensitive malignant lesion detection and reducing relapses. Shapira looks forward to the development of 'theragnostic' nanovehicles that carry four major components: A selective targeting moiety, a diagnostic imaging aid for localisation of the malignant tumor and its metastases, a cytotoxic small molecule drug(s) or innovative therapeutic biological matter and a chemosensitising agent to neutralise drug resistance – the advent of 'quadrugnostic' nanomedicine.

24.5.2 New detection methods and diagnostic devices

Nanoparticle probes, nanocantilever, nanowire and nanotube arrays are the subject of intensive research and are expected to solve the problem of early detection in the future. Accurate localisation of tumors and their metastases, via., nanoparticles loaded with a diagnostic aid, could in future facilitate the harnessing of other therapies, such as radiotherapy, photodynamic therapy and surgery. Heller group has described the goal of research as 'the development of a cancer therapy monitoring/diagnostic platform device'. This would provide real-time monitoring of patient blood for cancer cells, cell derived nanoparticulates (such as high molecular weight DNA fragments) and carry out cancer-related genotyping, gene expression and immunochemical analysis.

24.5.3 New applications and new targets

Superparamagnetic iron oxide (Fig. 24.4) nanoparticles (crystalline magnetite structures coated with dextran and dextran derivatives) are promising candidates for a number of applications, including magnetic resonance imaging and drug delivery. Bharali and Mousa believe that a major potential application of these nanoparticles is the diagnosis and treatment of central nervous system (CNS) tumors, particularly if USPIOs (ultrasmall supermagnetic iron oxide particles) are used, as they can be utilised as intravascular contrast agents, as well as for cellular imaging. One USPIO already showing great promise is Combidex, which has been undergoing clinical trials for the detection of lymph node metastases.

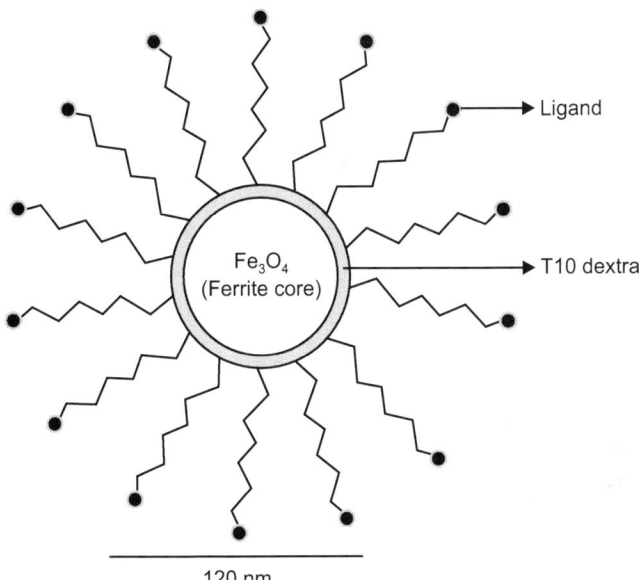

Figure **24.4**: Superparamagnetic iron oxide (Ferumoxtran-10).

Talekar and others predict that multifunctional superparamagnetic nano-carriers, with FR (folate receptor) targeting and pH mediated drug release, can be developed to achieve a decrease in tumor volume, as well as improved MRI sensitivity and decreased adverse effects. Metal coordination complexes also offer a diversity of formulations and the prospect of mechanisms that differ from those of organic drugs, including ligand substitution and metal-and ligand-centered redox properties. Therapeutic and imaging nanoparticles have normally used passive targeting to date, but active targeting needs to be used and further developed if drugs are to be delivered to specific classes of cells and specific intracellular sites in cancer cells.

24.5.4   Reducing the side effects

Nanoparticle encapsulation has already been shown to reduce unwanted accumulation of platinum in the kidneys from the platinum prodrug mitaplatin and reports show that metallodrugs loaded in nanoparticles cause less damage than the drugs on their own. So these formulations are predicted to be in line for further research and exploitation in the near future.

24.6   Bench to bedside: Translational perspectives

Nanotechnology has raised as many questions as it has answered and spawned new and unpredicted fields. With its vast array of potential applications in so many fields of science and industry, it is a prime candidate for multidisciplinary collaboration and the urgent need to see laboratory breakthroughs translated to clinical successes is increasingly recognised. Biochemists are increasingly working with scientists from fields not usually associated with medicine and the NIH's Nanomedicine Development Centers are staffed by multidisciplinary research teams, including biologists, physicians, mathematicians, engineers and computer scientists, whose first task has been to research the chemical and physical properties of nanoscale biological structures. Baseline work of this sort is vital if clinicians are to have the knowledge to develop new therapies.

Collins and others claim that translational research has been 'a powerful process that drives the clinical research engine', but feel that a stronger research infrastructure is needed in future to 'strengthen and accelerate this critical part of the clinical research enterprise.

To sum up, nanotherapeutics have already yielded significant breakthroughs in the detection, diagnosis and treatment of cancer and appear to have the potential to yield many more, with extensive and focused routes of research planned for the future and the possibility of nanotechnology-based cancer prevention. But it is clear that nanotechnology must be thoroughly understood and its risks assessed if it is to be developed safely and that the expertise of researchers in many fields needs to be brought together to move new discoveries out of the laboratory and into the clinical environment where patients can reap the benefits.'

Nanotechnology in medical diagnosis

25.1   Introduction

Nanotechnology is the creation and utilisation of materials, devices and systems through the control of matter on the nanometer (1 billionth of a meter)-length scale and its application in life sciences is termed as 'nano-biotechnology'. Nanotechnology is often represented by two fundamentally different approaches: 'top-down' and 'bottom-up'. 'Topdown' refers to making nanoscale structures by machining, templating and lithographic techniques, whereas 'bottom-up', or molecular nanotechnology, applies to building organic and inorganic materials into defined structures, atom-by-atom or molecule by-molecule, often by self-assembly or self-organisation. Biologists/chemists are involved in the synthesis of inorganic, organic and hybrid nanomaterials for the use in nanodevices, the development of novel nanoanalytical techniques.

Nanodiagnostics technologies like nanoscale visualisation, nanoparticle biolabels, biochips/microarrays, nanoparticle-based nucleic acid diagnostics, nanoproteomic-based diagnostics, biobarcode assays, nanopore technology, DNA nanomachines, nanoparticle-based immunoassays, nanobiosensors etc., are growingly popular in the field of medical diagnostics.

25.2   Nanotechnology-based biochips and microarrays

Generally the cell constituents exists in nano size and also the technology which is used to monitor/diagnose these biological molecules also falls in nano scale dimension which can be regarded as 'nanotechnology on a chip' is a new paradigm for total chemical analysis systems. Some examples of devices that incorporate nanotechnology-based biochips and microarrays are nanofluidic arrays and protein nanobiochips. These chips can be designed to interact with cellular constituents with higher specificity. One of the more promising uses of nanofluidic devices is isolation and analysis of individual biomolecules, such as DNA. This capability could lead to new detection schemes for cancer. One such device entails the construction of silicon nanowires on a substrate, or chip, using standard photolithographic and etching techniques, followed by a chemical oxidation step that converts the nanowires into hollow nanotubes. With this process the investigators can reliably create nanotubes with diameters as small as 10 nm, although devices used for biomolecule isolation contain nanotubes with diameters of 50 nm. Trapping DNA molecules requires a device consisting of a silicon nanotube connecting 2 parallel microfluidic channels. Electrodes provide a source of current used to drive DNA into the nanotubes. Each time a single DNA molecule moves into the nanotube, the electrical current suddenly changes. The current returns to its baseline value when the DNA molecule exits the nanotube. Nanofluidic technology is expected to have broad applications in systems biology, personalised medicine, pathogen detection, drug development and clinical research.

25.3   Protein microarrays/chips

Generally 2-D gel electrophoresis and mass-spectrometric methods (sensitivity to femtomole-attomole protein concentration) are used as the central tools for protein identification and characterisation between healthy and diseased samples. In addition to the above methods, protein microarray offer a powerful tool for screening thousands of proteins at a time, where variety of proteins such as antibodies and enzymes are immobilised in an array format on glass slide by robots. The surface of the glass slide is then probed with sample of interest that binds to relevant antibodies on chip which will be analysed by relevant detection method on array. Refinement in microarrays miniaturisation (with the advent of nanotechnology) will further contribute to molecular diagnostics and the development of personalised medicine.

Profiling proteins on arrays will be used in distinguishing the proteins of normal cells from early stage cancer cells and malignant metastatic cancer cells. It is also applicable to study the protein-protein interaction, the functional activity of proteins and in the discovery of disease markers and diagnosis. It can also be used for profiling of the serum to differentiate patients with pancreatic cancer from those with other pancreatic diseases and from healthy control.

Heart failure, arising from systemic disease or specific heart muscle disease, is one of the leading causes of morbidity and mortality in developing countries. Proteomics based approach in characterising overall changes in protein expression in heart diseases may provide new insights into the cellular involvement of interactions. Replacement of growing blood vessels in heart attack will be an addition to nanotechnology. The endothelial cells harvested in silicon molds shapes like capillaries; these cultivars would be delivered to the heart, via., microscopic machines called 'Angiochips' which will repair the damage caused by heart attack. Thus, protein microarrays offer the possibility of developing a rapid global analysis of the entire proteome, leading to protein-based diagnostics and therapeutics. It has been successfully applied to the study of complex trait, i.e., cardiovascular diseases, cancer and type II diabetes. Recent development in microarrays has come with technique called 'immunosensing', where patterned microarrays of antibodies to specific bacteria were used to perform a series of bacterial immunoassay and characterised using scanning probe microscopy

Protein chips, based on protein-binding silica-nanoparticles, are in developmental stage. The surface of a silica-nanoparticle with a diameter of less than 100 nm can be configured with many different capture proteins. The particles configured in this way are then applied to silicon carriers in thin, even layers. After contact is made with a sample, the chips can be analysed using state-of-the art mass spectrometry (MS)–Matrix-Assisted Laser Desorption/Ionisation (MALDI)–time-of-flight (TOF) MS.

25.4   Nanobiosensors

Biosensors are chemical sensors, in which recognition processes rely on biochemical mechanisms utilisation. They consist of a biological element (responsible for sampling) and a physical element (often called transducer, transmitting sampling results for further processing). Nanomaterials are exquisitely sensitive chemical and biological sensors. Ability to identify a particular type of cells or areas in a body makes the nanobiosensors to find its place in medical diagnostics. Based on the differences in volume, concentration, displacement and velocity, gravitational, electrical and magnetic forces, pressure, or temperature of cells in a body, nanosensors may be able to distinguish between and recognise certain cells, most notably those of cancer, at the molecular level in order to deliver medicine or monitor development to specific places in the body. In addition, they may be able to detect macroscopic variations from outside the body and communicate these changes to other nanoproducts working within the body.

One example of nanosensors involves using the fluorescence properties of cadmium selenide quantum dots as sensors to uncover tumours within the body. By injecting a body with these quantum dots, a doctor could see where a tumour or cancer cell was by finding the injected quantum dots, an easy process because of their fluorescence. Developed nanosensor quantum dots would be specifically constructed to find only the particular cell for which the body was at risk. A downside to the cadmium selenide dots, however, is that they are highly toxic to the body. As a result, researchers are working on developing alternate dots made out of a different, less toxic material while still retaining some of the fluorescence properties. In particular, they have been investigating the particular benefits of zinc sulphide quantum dots which, though they are not quite as fluorescent as cadmium selenide, can be augmented with other metals including manganese and various lanthanide elements.

In addition, these newer quantum dots become more fluorescent when they bond to their target cells. Quantum potential predicted functions may also include sensors used to detect specific DNA in order to recognise explicit genetic defects, especially for individuals at high-risk and implanted sensors that can automatically detect glucose levels for diabetic subjects more simply than current detectors. DNA can also serve as sacrificial layer for manufacturing CMOS IC, integrating a nanodevice with sensing capabilities. Therefore, using proteomic patterns and new hybrid materials, nanobiosensors can also be used to enable components configured into a hybrid semiconductor substrate as part of the circuit assembly. The development and miniaturisation of nanobiosensors should provide interesting new opportunities.

Other projected products most commonly involve using nanosensors to build smaller integrated circuits, as well as incorporating them into various other commodities made using other forms of nanotechnology for use in a variety of situations including transportation, communication, improvements in structural integrity and robotics. Nanosensors may also eventually be valuable as more accurate monitors of material states for use in systems where size and weight are constrained, such as in satellites and other aeronautic machines.

25.5   Nanoscale single-cell or molecule identification

Nanotechnology has facilitated the development of methods for detection of single cells or a few molecules. Nanolaser scanning confocal spectroscopy, with the capability of single-cell resolution, can be used to identify previously unknown properties of certain cancer cells that distinguish them from closely related nonpathogenic cells. Nanoproteomics, the application of nanobiotechnology to proteomics, can enable detection of a single molecule of protein. Biobarcode assays enable detection in body fluids of miniscule amounts of proteins that cannot be detected by conventional methods. A 2-dimensional method for mass spectrometry in solution is based on the interaction between a nanometer-scale pore and analytes. An applied electric current is used to force charged molecules (such as single-stranded DNA) one at a time into the nanopore, which is only 1.5 nm at its smallest point. As the molecules pass through the channel, the current flow is reduced in proportion to the size of each individual chain, allowing its mass to be easily derived. This single-molecule analysis technique could prove useful for the real-time characterisation of biomarkers.

25.6   Application of nanoparticles for discovery of biomarkers

Currently available molecular diagnostic technologies have been used to detect biomarkers of various diseases. Nanotechnology has refined the detection of biomarkers. Some biomarkers also form the basis of innovative molecular diagnostic tests. The physico-chemical characteristics and high surface areas of nanoparticles make them ideal candidates for developing biomarker-harvesting platforms. Given the variety of nanoparticle technologies that are available, it is feasible to tailor nanoparticle surfaces to selectively bind a subset of biomarkers and sequester them for later study using high-sensitivity proteomic tests. Biomarker harvesting is an underutilised application of nanoparticle technology and is likely to undergo substantial growth. Functional polymer-coated nanoparticles can be used for quick detection of biomarkers and DNA separation.

25.6.1 Nanoparticles for molecular diagnostics

Several nanoparticles have been used for diagnostics. Of these, the most frequently used are gold nanoparticles, Quantum dots (QDs) and magnetic nanoparticles.

25.6.2 Gold nanoparticles for diagnostics

Small pieces of DNA can be attached to gold particles no larger than 13 nm in diameter. The gold nanoparticles assemble onto a sensor surface only in the presence of a complementary target. If a patterned sensor surface of multiple DNA strands is used, the technique can detect millions of different DNA sequences simultaneously. The current nonoptimised detection limit of this method is 20 fmol/L. Gold nanoparticles are particularly good labels for sensors because a variety of analytical techniques can be used to detect them.

25.6.3 Quantum dots (QDs)

QDs are inorganic fluorophores that offer significant advantages over conventionally used fluorescent markers. Inorganic crystals of CdSe (cadmium selenide 200–10000 atoms wide), coated with ZnS (zinc sulphide). They emit fluorescent light when irradiated with low energy light. The size of the dots (< 10 nm) determines the frequency of light emitted (i.e., colour). The dots usually have a polymer coating with multivalent bio-conjugate attached, or are embedded into microbeads. Collection of dots of different size embedded to a given microbead emits distinct spectrum of colours - spectral bar code specific for this bead. Detection technique with the use of 10 intensity levels and 6 colours could theoretically provide 106 distinct codes. Quantum dots, for example CdSe-ZnS nanocrystals, do no emit in the near infrared, so they cannot be used for analysis in blood.

They have high sensitivity, broad excitation spectra, stable fluorescence with simple excitation and no need for lasers. Their red/infrared colours enable whole blood assays. QDs have a wide range of applications for molecular diagnostics and genotyping. QDs also enable multiplexed diagnostics and integration of diagnostics with therapeutics. The most important potential applications of QDs are for cancer diagnosis. Luminescent and stable QD bioconjugates enable visualisation of cancer cells in living animals. QDs can be combined with fluorescence microscopy to follow cells at high resolution in living animals. QDs have been coated with a polyacrylate cap and covalently linked to antibodies for immunofluorescent labelling of breast cancer marker Her2. Carbohydrate-encapsulated QDs with detectable luminescent properties are useful for imaging of cancer. QDs can be used in multicolour optical coding for biological assays.

25.6.4 Nanobarcodes

Freestanding, cylindrical nanoparticles with specific patterns of submicron stripes of noble metal ions, produced by alternating electrochemical reduction of the appropriate metals with the dimensions between 12 nm and 15 μm in width and 1–50 μm in length. The differential reflectivity of the adjacent metals provides contrast black and white stripes which makes them distinctive (like conventional barcodes) under light, or fluorescent microscopy, or mass spectrometry.

Nanobarcodes are useful in SNP mapping, Coding in multiplexed assays for proteomics, population diagnostics and in point-of-care handheld devices. Proteins detection by either mass spectrometry or fluorescence measure (after proteins immobilisation on a metal surface).

25.6.5 Magnetic nanoparticles

Iron nanoparticles, 15–20 nm in size and having saturation magnetisation, have been synthesised and embedded in copolymer beads of styrene and glycidyl methacrylate (GMA), which were coated with polyGMA by seed polymerisation. The resulting Fe/St-GMA/GMA beads had diameters of 100–200 nm. Coating with polyGMA changed the zeta potential of the beads from 93.7 to 54.8 mV, as measured by an electrophoresis method. As revealed by gel electrophoresis, this process facilitates nonspecific protein adsorption suppression, which is a requisite for nanoparticles to be applied to carriers for bioscreening. Nanoparticles are used as labelling molecules for bioscreening.

Super paramagnetic nanoparticles are useful for cell-tracking cells and for calcium sensing. Ferrofluids consist of a magnetic core surrounded by a polymeric layer coated with antibodies for capturing cells. A family of calcium indicators for MRI is formed by combining a powerful SPIO nanoparticle-based contrast mechanism with the versatile calcium sensing protein calmodulin and its targets. Super paramagnetic nanoparticles measuring 2–3 nm have been used in conjunction with MRI to reveal small and otherwise undetectable lymph-node metastases. Ultra small SPIO enhances MRI for imaging cerebral ischemic lesions. A dextran-coated iron oxide nanoparticle enhances MRI visualisation of intracranial tumours for more than 24 hr.

25.7 Nanoparticles as biosensors components

There are several kinds of nanoparticles that can be used as biosensors components. Most of them work as probes recognising and differentiating an analyte of interest for diagnostic and screening purposes. In such applications biological molecular species are attached to the nanoparticles through a proprietary modification procedure. The probes are used then to bind and signal the presence of a target in a sample by their colour, mass, or other physical properties. The molecular binding is a subject of the biological surface science, which is closely related to the research on modification of nanostructures properties by controlling their structure and their surface at a nanoscale level. Both fields cover broad area, in which one can locate the use of nanostructured platinum-lipid biolayer composite as biosensor or research on endothelisation and adherence of the cells to nanostructure surfaces. Biosensors based on quantum dots, nanobarcodes, metallic nanobeads, silica nanoparticles, magnetic beads and carbon nanotubes can be qualified to the group of nanoprobes.

The other biosensors employ nanoparticles in a different way. They work as sieves through which charged molecules are transported in an electrical field. Particles with engineered nanopores are used. But solid-state materials are not the only resources for nanoparticles construction. For example DNA-nanopores made of α-haemolysin protein channel mounted in a lipid bilayer with ~2 nm in diameter are made of organic material. These nanopores are able to discriminate between individual DNA strands up to 30 nucleotides in length, differing by a single base substitution. There are several articles concerning this subject, while a comprehensive review is provided by. Nanopores discuss native α- haemolysin sensor applications and present sensors based on modified α- haemolysin (together with description of other organic pores and synthetic nanopores, supporting structures and applications).

25.8   Application of nanodiagnostics in infectious diseases

The rapid and sensitive detection of pathogenic bacteria at the point of care is extremely important. Limitations of most of the conventional diagnostic methods are lack of ultra sensitivity and delay in getting results. A bioconjugated nano-particle-based bioassay for *in situ* pathogen quantification can detect a single bacterium within 20-min. Detection of single-molecule hybridisation has been achieved by a hybridisation-detection method using multicolour oligonucleotide-functionalised QDs as nanoprobes. In the presence of various target sequences, combinatorial self assembly of the nanoprobes, via., independent hybridisation reactions leads to the generation of discernible sequence specific spectral codings. This method can be used for genetic analysis of anthrax pathogenicity by simultaneous detection of multiple relevant sequences. A spectroscopic assay based on SERS using silver nanorods, which significantly amplify the signal, has been developed for rapid detection of trace levels of viruses with a high degree of sensitivity and specificity. The technique measures the change in frequency of a near- infrared laser as it scatters viral DNA or RNA. That change in frequency is as distinct as a fingerprint. This novel SERS assay can detect spectral differences between viruses, viral strains and viruses with gene deletions in biological media. The method provides rapid diagnostics (60 s) for detection and characterisation of viruses generating reproducible spectra without viral manipulation. This method is also inexpensive and easily reproducible.

25.9 Future issues in the development of nanobiosensors

New biosensors and biosensor arrays are being developed using new materials, nanomaterials and microfabricated materials and new methods of patterning. Biosensor components will use nanofabrication technologies. Nano sized devices can be produced by use of nanotubes, fullerenes (buckyballs) and silica and its derivatives. Some of the challenges will be development of real-time noninvasive technologies that can be applied to detection and quantification of biological fluids without the need for multiple calibrations using clinical samples. It would be desirable to develop multiple integrated biosensor systems that use doped oxides, polymers, enzymes, or other components to give the system the required specificity. Such integrated sensor systems would include all of the sensor components, software, plumbing and reagents along with sample processing. There is also a need for reliable fluid handling systems for 'dirty' fluids and for relatively small quantities of fluids (nanoliter to attoliter quantities). These should be low cost, disposable, reliable and easy to use as part of an integrated sensor system. Sensing in pico liter to atto liter volumes might create new problems in the development of microreactors for sensing and novel phenomena in very small channels.

25.10 Safety issues of nanoparticles for diagnostics

Potential toxic effects are a concern with *in vivo* use of nanoparticles but not with *in vitro* diagnostics, which forms the major portion of laboratory diagnostics. There are environmental concerns about the release of nanoparticles during manufacturing of nanoparticles and the environmental effects. These are being studied along with naturally present nanoparticles in the atmosphere. There are still many unanswered questions about the fate of nanoparticles introduced into the living body. Because of the huge diversity of materials used and the wide range in size of nanoparticles, these effects will vary considerably. QDs made with fluorescent labels of calcium selenide or zinc sulphide to increase stability may release potentially toxic cadmium and zinc ions into cells.

A high-throughput gene expression test determined that specially coated QD fluorescent nanoprobes affect only 0.2% of the human genome, dispelling the concern that the mere presence of these potentially toxic sentinels disrupts cell function. It is conceivable that particular sizes of some materials may have a bearing on toxic effects. Concern centers around nanoparticles smaller than 20 nm in diameter, which can penetrate the cells. One limitation for the approval of *in vivo* nanomaterials for human diagnostics would be that demonstration of safety of nanoparticles would be required. A number of studies have been done, but at this stage, no conclusions can be drawn about the safety of nanoparticles.

25.11   Future prospects of nanotechnology

Medical diagnosis, proper and efficient delivery of pharmaceuticals and development of artificial cells are the medical fields where nanosize materials have found practical implementations. The application of nanotechnology to medicine, nanomedicine, subsumes three mutually overlapping and progressively more powerful molecular technologies.

First, nanoscale-structured materials and devices that can be fabricated today hold great promise for advanced diagnostics and biosensors, targeted drug delivery and smart drugs and immunoisolation therapies.

Second, biotechnology offers the benefits of molecular medicine, via., genomics, proteomics and artificial engineered microbes.

Third, in the longer term, molecular machine systems and medical nanorobots will allow instant pathogen diagnosis and extermination, chromosome replacement and individual cell surgery *in vivo* and the efficient augmentation and improvement of natural physiological function. Within the next decade, measurement devices based on nanotechnology, which can make thousands of measurements very rapidly and very inexpensively, will become available.

Future trends in diagnostics will continue in miniaturisation of biochip technology to the nanoscale range. Molecular electronics and nanoscale chemical sensors will enable the construction of microscopic sensors capable of detecting patterns of chemicals in a fluid. Estimates of plausible device capabilities have been used to evaluate their performance for typical chemicals released into the blood by tissues in response to localised injury or infection. These indicate that the devices can readily enable differentiation of a single cell–sized chemical source from the background chemical concentration *in vivo*, providing high-resolution sensing in both time and space. With currently used methods for blood analysis, such a chemical source would be difficult to distinguish from background when diluted throughout the blood volume and withdrawn as a blood sample. The trend will be to build the diagnostic devices from the bottom up, starting with the smallest building blocks. Whether interest and application of nanomechanical detection will hold in the long range remains to be seen. Another trend is to move away from fluorescent labelling as miniaturisation reduces the signal intensity, but there have been some improvements making fluorescent labelling methods viable with nanoparticles.

To sum up, over coming years nanotechnology can be practically applied in creation of artificial cells, tissues and organs. Artificial cells are being actively investigated for use in the replacement of defective or incorrectly functioning cells and organs, especially related to metabolic functions. Nanotechnologies promise to extend the limits of current molecular diagnostics and enable point-of-care diagnosis, integration of diagnostics with therapeutics and development of personalised medicine. The most important clinical applications of currently available nanotechnology are in the areas of DNA detection assay, biomarker discovery, cancer diagnosis and detection of infectious micro-organisms. Nanomedicine promises to play an important role in the future development of diagnostic and therapeutic methods.

Section V

Molecular biotechnology and society

Societal issues of biotechnology

26.1   Introduction

Advances in biotechnology and their applications are most frequently associated with controversies. On their perception to biotechnology, the people may be grouped into three broad categories:

1. Strong opponents who oppose the new technology, as it will give rise to problems, issues and concerns humans have never faced before. They consider biotechnology as an unnatural manipulative technology.
2. Strong proponents who consider that the biotechnology will provide untold benefits to society. They argue that for centuries the society has safely used the products and processes of biotechnology.
3. A neutral group of people who have a balanced approach to biotechnology. This group believes that research on biotechnology (with regulatory systems) and extending its fruits to the society should be pursued with a cautious approach. The risks and benefits of the developments of biotechnology may not be much different from that of any other branch of science.

26.2   Benefits of biotechnology

The fruits of biotechnology are beneficial to the fields of healthcare, agriculture, food production, manufacture of industrial enzymes and appropriate environmental management. It is a fact that modern technology in various forms is woven tightly into the fabric of our lives. Our day-to-day life is inseparable from technology. Imagine life about 1–2 centuries ago where there was no electricity, no running water and sewage in the streets, unpredictable food supply and an expected life span of less than 40 years. Undoubtedly, technology has largely contributed to the present day world we live in. Many people consider biotechnology as a technology that will improve the quality of life in every country, besides maintaining living standards at a reasonably higher level. The probable positive and negative effects of biotechnology, with special reference to developing countries are given in Table 26.1.

26.2.1   ELSI of biotechnology

Why so much uproar and negativity to biotechnology? This is mainly because the major part of the modern biotechnology deals with genetic manipulations.

Table 26.1: The probable positive and negative effects of biotechnology (particularly In the developing countries).

Positive effects
Improved health and life span (through more effective pharmaceuticals, vaccines, etc.)
Increased crop production and more food
Decreased dependence on import of food, fertilisers and chemicals
Enhanced biomass production for energy
Use of plants and animals as bioreactors
Improved food storage facilities
Increased production and health of livestock
Negative effects
More dependence on developed countries and private companies for technology and resources
Increase in unemployment as the labour requirement is less
Reduction in natural biodiversity and natural ecosystems
More legal and financial complications
Increased growing of cash crops at the cost of food crops
Production of more and more herbicide resistant plants

These unnatural genetic manipulations, as many people fear, may lead to unknown consequences. Ethical, Legal and Social Implications (ELSI) broadly covers the relationship between biotechnology and society with particular reference to ethical and legal aspects.

Risks and ethics of biotechnology

The modern biotechnology deals with genetic manipulations of viruses, bacteria, plants, animals, fish and birds. Introduction of foreign genes into various organisms raises concerns about the safety, ethics and unforeseen consequences.

Some of the popular phrases used in the media while referring to experiments on recombinant DNA technology are listed:

1. Manipulation of life.

2. Playing with God.

3. Man-made evolution.

The major apprehension of genetic engineering is that through recombinant DNA experiments unique micro-organisms or viruses (either inadvertently or sometimes deliberately for the purpose of war) may be developed that would cause epidemics and environmental catastrophes. Due to these fears, the regulatory guidelines for research dealing with DNA manipulation were very stringent in the earlier years.

So far, risk assessment studies have failed to demonstrate any hazardous properties acquired by host cells/organisms due to transfer of DNA. Thus, the fears of genetic manipulations may be unfounded to a large extent. Consequently, there has been some relaxation in the regulatory guidelines for recombinant DNA research. It is now widely accepted that biotechnology is certainly beneficial to humans. But it should not cause problems of safety to people and environment and create unacceptable social, moral and ethical issues.

The public fears of biotechnology, besides some of the risks, social and ethical issues can be better understood with special reference to the following:

1. Therapeutic products for use in healthcare.
2. Genetic modifications of foods and food ingredients and their consumption.
3. Release of genetically engineered organisms into the environment.
4. Applications of human genetic research.

Recombinant therapeutic products for human healthcare

It is fortunate that there is no serious criticism about the use recombinant products for medical applications. This is mostly because the therapeutic products and strategies are designed to cure diseases, alleviate sufferings and improve the quality of life. Further, the products are used under the medical supervision. In general, the recombinant products designed for human healthcare are more readily acceptable by the public. Good examples are the use of insulin, interferons, tissue plasminogen activator and various vaccines

26.2.2 Genetic modifications and food consumption

The overall objectives of genetic modifications with reference to foods are listed:

1. To increase the quality and quantity of existing foods.
2. To produce new products.
3. To improve the financial returns.

Each country has its own regulations for introducing foods into the market. For instance, in U.S., the Food and Drug Administration (FDA) regulates the introduction of foods, drugs, pharmaceuticals and medical devices into the marketplace. As regards the foods and food ingredients developed by genetic engineering, the FDA believes that no new regulations are needed. The existing regulations for the assessment foods for safety by toxicity, allergen city and impurity testing are adequate.

If the chemical composition of the existing food is altered by genetic modification, it should be specified and the new product should be accordingly labelled. To highlight the public perception of genetically modified foods, some selected examples of food ingredients and their acceptance, or certain controversies related to their use, are briefly described.

Chymosin

Chymosin is milk clotting proteolytic enzyme that hydrolyses the milk protein casein to produce curd, which in turn is processed into cheese. Traditionally, chymosin is derived from the stomach of calves in the form of rennet. By genetic engineering techniques, the chymosin gene was cloned and expressed in *E. coli*. This resulted in a large scale and cost- effective production of chymosin. The chemical composition, structure and biological activity of recombinant chymosin were identical to the chymosin of rennet. FDA gave license to chymosin for its commercial use and it is now widely used in cheese making.

Tryptophan

In 1989–1990, an unusually high incidence of the disease eosinophilia-myalgia syndrome (EMS) was reported in U.S. EMS is a rare disease with muscular pain and respiratory complications and may be fatal. Investigations revealed that the victims of EMS were consuming large quantities of food supplement tryptophan (obtained from one company).

This tryptophan was produced by genetically engineered micro-organisms. Chemical analysis of the commercial preparations revealed the presence of certain metabolic derivative of tryptophan. The most important among them was 1-1′ ethylene—*bis* (tryptophan) which was responsible for EMS (based on the results of animal experiments).

It was concluded that the pharmaceutical company did not take adequate care for purification of tryptophan. Consequently, recombinant tryptophan (even without the impurities) was banned for human consumption in U.S.

Bovine somatotropin story

Bovine somatotropin (BST, also known as bovine growth hormone), when injected to dairy cattle, increases the milk production significantly. By recombinant DNA technology, the gene for BST was cloned and expressed in *E. coli*. The recombinant BST (rBST) so produced, when injected to cows was found to increase milk production by 20–25%.

The effect of rBST was considered from two aspects—on the animals and consumers:

1. Effect of rBST on animals: Administration of rBST can produce localised swelling at the site of injection. There may be some other adverse effects like increased susceptibility to infection, decreased reproductive capability. The proponents of rBST argue that these problem could occur even in normal animals.

2. Effect on human health: The natural BST or even rBST increases the body levels of insulin-like growth factor-l (IGF-I) which in turn enhances milk production. There is evidence that IGF-I stimulate growth of cancer

cells. This causes concern among the consumers of milk produced by using rBST.

The counter argument is that rBST injection, after about 100 days of lactation begins, is not associated with increased levels of IGF-I. The opponents of rBST strongly feel that since milk is consumed by most people, any inherent risk, however small is unacceptable. Recombinant BST was licensed in U.S., for use in dairly cattle in 1994. Some countries in fact have banned the use of rBST. There are some people (particularly among the scientists) who believe that the hue and cry raised against rBST is more due to economic and political reasons. It is feared that by rBST use, the dairy industry may be controlled by large industrial groups and the small dairy farms may become unprofitable. The rBST story is an interesting illustration of the problems surrounding the use of genetic engineering.

26.2.3   Recombinant foods and religious beliefs

Some of the ethical concerns of the use of recombinant foods are related to religious beliefs, besides food habits. For instance:

1. Transfer of pig genes into sheep may offend the sentiments of Jews and Muslims.
2. Introduction of animal genes into food plants may invite opposition by strict vegetarians.
3. Transfer of human genes to food animals may be unacceptable to some people.
4. Feeding of human gene — containing organisms to animals sounds in bad taste (at present, the genetically modified yeast that produce recombinant proteins after their use, are fed to animals).

The religious groups, in general, are selective about the foods to be eaten. However, they are not so rigid when it comes to the use of medically-derived products. For instance — Jews and Muslims may accept pig-derived insulin for use in diabetic patients. This is due to the fact that all religious faiths consider human life is the most valuable and its preservation the first priority. Further, the general belief is that the human body is violated only by oral consumption and not by injection or surgical interventions.

Eating genes every day

Every day, we eat plants and animals and various products derived from them, besides a large number of bacteria. In other words, we regularly consume genetic material, the DNA, organised into genes in various organisms. So far no one has attempted to categorise genes as vegetarian and non-vegetarian, as we do for foods.

Are GM foods safe?

The production of transgenic plants and animals by genetic engineering techniques has now become routine. These organisms will enter the food chain in the form of genetically modified foods (GM foods). Some social and environmental groups are against the consumption of GM foods. These people insist that the GM foods should be specifically labelled. As regards the safety of GM foods, opinions range from one extreme that they are absolutely safe and improve human nutrition to the other that they should not be consumed at all. Most of the people have opinions somewhere between these two extremes.

26.2.4 Release of genetically engineered organisms

The release of genetically engineered organisms (GEOs), also called as genetically manipulated organisms (GMOs) into the environment has been a controversial issue and continues to be so. It is feared that the release of GEOs into the environment could have far-reaching consequences.

This is due to the fact that the living GEOs proliferate, persist, disperse and sometimes may transfer their DNA into other organisms. It is further feared that there exists a possibility of GEOs displacing the existing organisms, besides creating new species. This may lead to severe environmental damage. For the reasons stated above, the regulatory authorises are very careful in permitting the field trails of GEOs. Further, the release of GEOs into the environment has to be carefully monitored and recorded.

Ice-minus Pseudomonas syringae

A genetically modified strain of *Pseudomonas syringae* was the first GEO that was given permission for field trails in 1987. This organism is a genetically engineered ice-minus strain, when sprayed onto the leaves could prevent frost damage to the plants.

Field trails with other GEOs

A number of open-field trails have been conducted with several GEOs during the past two decades or so. The studies concluded that the genetically modified micro-organisms do not persist in environment for long, do not transfer the genes into other organism and do not exhibit any abnormal biological functions. Thus, the initial apprehensions on the use of GEOs appear to be unfounded.

Release of transgenic plants and animals

The transgenic plants developed for higher quality and quantity of foods, in general, are more liberally permitted to go to fields. It is generally believed that the transgenic plants do not significantly differ from the natural cultivars (traditional plants) obtained by plant breeding experiments.

The transgenic *Bt*-plants such as cotton, corn, soyabean and potato were approved for cultivation in U.S. However, some countries did not allow *Bt*-plants in their fields, e.g., *Bt*-rice was not allowed in Philippines, *Bt*-cotton in France. The transgenic animals have not posed as big a problem as the transgenic plants. This is due to the fact that animals can be much easily identified and contained (since copulation in animals can be done as desired, unlike in plants where pollination is difficult to control). A majority of transgenic animals are used for medical purposes, hence they are generally appreciated. But the major problem for transgenic animals comes from animal activists. Many people and also some governments, are suspicious of the use of GEOs due to various reasons-risks, societal beliefs and economic concerns.

Biological warfare

The most serious hazard associated with genetic engineering is the construction of harmful biological agents (viruses, micro-organisms) either deliberately or otherwise. However, so far there have been no records of any new infectious agents created by recombinant DNA technology.

Most of the countries of the world are signatories to the Biological Weapons Conventions of 1972. As a signatory, a nation pledges 'never to produce microbial or other biological agents, or toxins, whatever may be their method of production, for use in wars'. Many people are, however concerned about the possible use of gene manipulations for military purposes.

26.2.5   Applications of human genetic rDNA research

The ultimate goal of advances in biotechnology is for the benefit of mankind (either direct or indirect). Biotechnology largely contributes to human genetic research involving the following areas:

1. Genetic testing and screening for diseases.
2. Genetic portfolios.
3. Human gene therapy.

Genetic testing and screening

Techniques are now available for prenatal testing to specifically detect whether a fetus carries genetic defects. This will help the parents to be better prepared for the future baby. The negative aspect of prenatal testing is that the couple may opt for abortion even for a minor genetic defect or sometimes for gender bias.

Genetic portfolios

The elucidation of the entire human genome sequence and identification of genes has now become a reality. It may soon be possible to have individual genetic portfolios that will diagnose future health complications, e.g., risk for cancer, heart disease. The genetic portfolios (based on the genes) will foretell the individuals future which is now being predicted through stars (astrology).

Genetic portfolios of individual may pose certain problems with regard to marriages, insurances. Who would like to be a spouse of someone who will soon be a victim of cancer or heart attack? Which insurance company would insure a person with a very high risk of diseases? Many ethical committees are of the opinion that insurance companies should not require, or should not be allowed to have access to individuals' genetic portfolios.

Human gene therapy

Theoretically, correction of genetic defects is possible by gene therapy. The present status of human gene therapy has been described. From the ethical perspective, gene therapy involving introduction of genes into a patient is comparable to the practice of transplantation of organs (e.g., heart, liver, lungs).

Therefore, there is not much controversy over gene therapy, as long as it is intended to be used to alleviate serious medical disorders. However, the gene therapy must be under a close supervision to satisfy medical, legal, ethical and safety implications, besides addressing the public concerns.

Germ line gene therapy

This is not being carried out at present due to technical, ethical and social reasons. Manipulation of germ cells will lead to serious problems and complications. At one international meeting on biotechnology the following was the final message given to biotechnology companies on the applications of genetic engineering to humans. 'Provide the information and listen to the public'.

26.2.6   Human embryonic stem cell research

The human embryonic stem cell (ESC) lines were established in November 1998. These cells are capable of giving rise to any human cell type. ESC lines open the possibilities of treating diseases with cell therapy. Disorders that involve the loss of normal cells such as diabetes mellitus, Parkinson's disease, Alzheimer's disease could perhaps be corrected with cell therapy. This may however, take more time. There are many ethical and legal issues involved in ESC research, besides several objections raised by the public. In fact, the U.S. Federal Government has banned the use of federal funds for human embryo research for over 30 years. At present, most of research on ESC is being supported by private companies.

26.2.7   Cloning humans

After the cloning of the sheep Dolly, some groups of researchers naturally became interested to explore the possibilities of cloning humans. The very thought of human cloning has become a highly charged and controversial issue. As such, considering the biomedical ethics, most of the countries have banned research related to human cloning. It cannot be predicted at present whether someday human cloning may become inevitable.

Scientists who were awarded Britain's first licence for human cloning have recently created an early stage of human cloned embryo by using nuclear transfer. The researchers, as such, are not interested to make babies by cloning. Instead, they wish to create test-tube embryos to supply stem cells that can give rise to every tissue in the body.

By this approach, it might be one day possible to repair tissue damages and cure many diseases, e.g., Parkinson's disease, diabetes mellitus.

26.2.8   Biotechnology and the developing countries

A major proportion of research related to biotechnology is carried out by the developed countries. The fruits of biotechnology are probably, more useful and relevant to developing countries, as illustrated with the following applications:

1. Increased levels of nutrition with improved nutrient composition in the foods (through transgenic plants).
2. Prevention of child deaths by appropriate immunisation (using recombinant vaccines).
3. Supply of clean drinking water and improved sewage disposal (by appropriate bio-technological treatments).

Despite the known benefits, some of the developing countries are reluctant to open doors for the advances made in biotechnology. This may be more due to political considerations rather the economic reasons. The applications of biotechnology, particularly to human healthcare, may someday be regarded as a barometer to evaluate the progress of a nation.

26.3   Social aspects of biotechnology

As stated, claims have been made by the public as well as by social scientists that there must be a more responsible way of governing, developing and applying new biotechnologies. Examining biotechnology from a societal perspective is often interpreted narrowly as bioethics. For example, the responsibility of the state could be seen as including the support of research and development of biotechnology so that new jobs can be created. The discussions about biotechnology, therefore, also touch economic, scientific, technical and legal aspects of the research, development and application of biotechnology. Different actors (like the state, industry, research community and the public) are also affected, but in varying degrees according to the problem under consideration. Public discussion ranges from the lack of venture capital for biotechnology firms to the ethical acceptability of prenatal screening and from the health risks of genetically modified organisms (GMOs) to informed consent of objects for biomedical research.

First, biotechnology is regarded as a unified branch of technology in many reports, policies and documents; and these documents form the other half of empirical data. A number of industrial associations, university institutes, etc., bear the word biotechnology in their name. Therefore, at least on a strategic and policy level, biotechnology is often considered as an entity.

Second, examining biotechnology as a whole reveals the differences inside biotechnology–for example, between medical (red) and environmental (green) biotechnology–but also doing so gives a more general idea of how responsibilities are linked to technologies. Buttel argues that there are a number of reasons why, in social science, red and green biotechnology have been dealt with separately and two distinct literatures have formed. For example, biological constraints in the commercialisation of agriculture and in medical biotechnology are distinct from each other. Another example is that the position of citizens tends to be framed with 'choice' in green biotechnology and with 'access' in red biotechnology. Buttel, however, calls for a more integrated approach in studying biotechnology. He names three areas where similarities and more general issues for social sciences can easily be detected: The role of patenting in biotechnology R&D, the importance of discursive framings and struggles in policymaking and public discussion and the timing of public input into decision-making. This study offers insights into the two latter areas.

Indeed, the responsibilities ascribed to different actors–researchers, industry, or citizens, for example–are one of those general aspects of biotechnology (or any other technology for that matter) that are not limited to specific technologies, sectors, or applications. The study of the notion of responsibility helps to describe, understand and criticise the system and structures of knowledge production and knowledge policies.

The empirical material for this research comes mainly from Finland and it is the Finnish research and development of biotechnology and the Finnish science and technology governance that are analysed. However, this is not a mere case study of Finland; rather it contributes to the international scientific discussion about the social aspects of biotechnology. Biotechnology is a highly international field where research, business, markets and legislation span national borders and Finnish biotechnology is part of the European and global structures and procedures. There are, of course, special characteristics in the Finnish biotechnology sector and its governance. Compared to many other European countries, Finnish people have adopted a very moderate stance towards genetic technologies and they trust the authorities and experts dealing with biotechnology. In addition, pressures to commercialise research and the introduction of 'market-logic' to the public sector have gained considerable ground in Finland. So it could be argued that in Finland, lack of public resistance and a general consensus among experts and decision-makers about the benefits of investing in biotechnology create a seemingly steady and fruitful environment for developing new biotechnology. This study will challenge the simplicity of this picture.

In addition, this Finnish case is reflected to the general developments in the European Union. European Union policies and documents serve as a fruitful comparison point as they are also, in practise, the reference points of Finnish policy making. Finnish legislation, guidelines and strategies need to correspond to those of the European Union. However, when European models are translated to the Finnish environment, they become contextualised differently; and both practises and discourses can change considerably from the originals.

The biotechnology industry is an evolving sector of this century having potential to significantly influence our lives. It has unidentified potentials to contribute to sustainable development, but is also feared by society for its possible known and unknown impacts.

Amplified transparency brought about by media, new concerns and expectations of society towards business as well as social criteria are evermore influencing peoples investment decisions, making corporate social responsibility of increasing importance to companies.

26.4 Special characteristics of biotechnology

The introduction and characterisation of biotechnology showed that this technology has unique features. Those features are the bases for the criteria developed to analyse the relevance of social aspects for biotechnology. For this reason, special emphasis is given to them in this section.

The characteristics of biotechnology can be summarised as follows:

1. Biotechnology, by the use of living organisms, has a considerable possibility to contribute to sustainable development and increase our quality of life.

2. Biotechnology has potentials for being applied in cleaner production (through dematerialisation and substitution), but is also applicable for End-of-life purposes.

3. Biotechnology is knowledge based and requires highly-skilled labour.

4. Biotechnology is new, developing and growing rapidly.

5. Because of the use of living organisms, biotechnology has a dynamic (and maybe uncontrollable) nature. The reason is that living things can change (randomly or if outside conditions change), which leads to an uncertainty regarding future impacts.

6. Due to the fact mentioned in point 5, biotechnology raises concerns among the public but also scientists (they are concerned about the technology and its potential impacts).
7. Biotechnology requires interdisciplinary teams to explore its potentials.
8. Biotechnology raises, to a certain extent, ethical questions, i.e., the application of genetic engineering.

26.5 Corporate social responsibility and social performance

As there are few regulative requirements regarding social reporting today many company's activities are based on voluntary actions. The concept of Corporate Social Responsibility (CSR) implies action on a voluntary basis, going beyond compliance and investing in social (human) capital and stakeholder relations.

According to the European Commission there are set of factors driving this recent move towards CSR:

1. New concerns and expectations from citizens, consumers, public authorities and investors in the context of globalisation and large scale industrial change.
2. Social criteria are increasingly influencing the investment decisions of individuals and institutions both as consumers and as investors.
3. Increased concern about the damage caused by economic activity to the environment.
4. Transparency of business activities brought about by the media and modern information and communication technologies.

Those factors show the need for businesses to communicate their social performance, be transparent and open about their activities and to respond to the expectations of stakeholders in order to achieve long-term acceptance and support. On the other hand, it is becoming increasingly difficult for businesses to address all those issues since it requires companies to open themselves in a way they might not feel comfortable with (e.g., the protection of business secrets becomes increasingly difficult).

However, the nature of social aspects is that they influence society and they could do so negatively. For this reason and with respect to the precautionary principle, openness and transparency should be required to avoid negative impacts in the first place. Innovative companies, who are pioneers regarding social responsibility, daring to experiment and take risks also in the field of openness, will likely be able to set standards for others to follow. These standards are usually later required by public and codified by governments in order to push followers to do the same.

26.5.1   Importance of corporate social responsibility

Because business is currently seen not as something different but as an element of society and its values, CSR plays an important role in business operations. The Millennium poll on CSR in 1999 revealed that out of 25000 interviewed people (from 23 countries on 6 continents) about half are paying attention to the social behaviour of companies. This shows that CSR is widely acknowledged as an important issue, but there is still little guiding help for enterprises of how to put the concept into practice.

The importance of stakeholder interests for the performance of organisations operating today is clearly indicated by this definition, since organisations more and more need to align their values according to the expectations expressed by society. This statement is supported by the Global Reporting Initiative (GRI), which perceives the social performance as a key ingredient in assuring an organisation's licence to operate.

The aim of the GRI is to give guidelines for sustainability reporting. CSR does not address sustainability itself directly, but is one vital part of sustainable development. Additionally, there are many overlapping social issues, such as community involvement, human, employee and stakeholder rights that are identified to be important when it comes to social performance. CSR is a social investment motivated by long-term thinking and hence contributes to sustainable development, which per definition not only aims at meeting needs of current but also those of future generations. However, needs will be defined, they for sure include claims on the social level. Based on this connection, a company can make a contribution to sustainable development by showing a socially responsible behaviour.

In addition to that, acting socially responsible is also in the interest of the company and makes good business sense. Having knowledge about external expectations and the position of the company related to that and using this knowledge as a strategic management tool can also lead to a successful performance in the economic dimension. For a long-term strategy, CSR should be seen more as an investment in the company rather than as a cost or burden. The European Commission points out that going beyond compliance through investing in social capital such as employee training and working conditions is very likely to improve the company's competitiveness.

Besides that, subscribing to the concept of CSR influences factors such as risk management and reputation. The values of a company are reflected in the way employees behave and this influences the perceived image of the company by the public (e.g., a single focus on profit might lead to mistrust and a loss of reputation, respect and loyalty).

26.5.2   Internal dimension of corporate social responsibility

The internal dimension basically covers social responsibility practices within the enterprise and involves employees. Issues such as health and safety, investing in human capital and managing change are dealt with from this perspective. The subject of health and safety is already addressed by legislation. Nevertheless, enterprises try to increase the quality of performance by implementing standards beyond legislation not only within the company but also require those from their contractors and suppliers.

By promoting this preventive approach, companies can gain a competitive advantage, because they are able to make financial savings. Those savings are realised, for example, through fewer absent hours of employees caused by accidents. Furthermore, companies proving high health and safety standards and as a result less risks could potentially have lower insurance premiums to pay than other companies.

The matter of human resource management includes several aspects that determine success. Especially in the sector of new technologies, the main resources for enterprises are highly-skilled and motivated employees. To ensure retention of those workers the enterprise should provide activities including *'lifelong learning, empowerment of employees, better information throughout the company, better balance between work, family and* leisure, greater work force diversity, equal pay and career prospects for women, profit-sharing and share ownership schemes and *concern for employability as well as job security'*. Besides that, non-discriminatory practices are likely to create commitment from employees (especially from minorities) to the enterprise.

The widespread restructuring going along with globalisation causes serious concerns among stakeholders. Involving affected parties in early discussion stages as well as building social inclusion partnerships and having a long-term labour strategy, can facilitate the adaptation to change and lessen the social impacts of restructuring.

26.5.3   External dimension of corporate social responsibility

CSR is not restricted to issues within a company but extends beyond the boundaries of enterprises and requires the involvement of all affected stakeholders. Besides employees and shareholders, the European Commission mentions business partners and suppliers, customers, public authorities and Non-governmental Organisations (NGOs) as additional stakeholders. Local communities are especially important to enterprises, as there is often a close interaction (and dependence) among them. The reputation at the company's location mainly influences the interest of local people to work as employees or to act as customers and therefore, affects the competitiveness of the company.

Regarding consumers, the design of products and services in a socially responsible manner needs to fulfil criteria of quality, reliability, safety and availability for as many people as possible. Socially responsible acting also positively impacts business-to-business relationships. The reliable delivery of high-quality products and long-term contracts with suppliers are just two examples of this. For larger companies, another practice to demonstrate social responsibility is the promotion of small start-up enterprises or local small- and medium-sized enterprises (SMEs), thus improving the economic situation of the region. Human rights are another very important external dimension of CSR, including, for example, the problem of corruption and child labour. The identification of the company's responsibilities is a challenging question especially because '*human rights are a very complex issue presenting political, legal and moral dilemmas*'.

Several legal requirements set minimum standards applicable to all enterprises. Nevertheless, to a greater extent, sectors and companies establish their own voluntary codes of conduct to demonstrate commitment to social concerns.

26.5.4   Social responsibility and economic benefits

One of the driving forces mentioned above by the European Commission is the influence social criteria have on investment decisions made by people. Since many companies are listed in the stock market those investment decisions can directly affect the economic success of a company. There are indices existing that especially value the commitment of companies towards a more sustainable development. Green Effects for example is an international, ethic-ecological fund investing in the Nature Share Index (NAI) (Natur-Aktien-Index). NAI is a measure of the economic success of enterprises contributing to an ecologically and socially sustainable development and consists of 20 international companies from different sectors that are estimated to be economically successful in the long run. The decision of which companies are represented in the NAI is based on binding performance criteria, including environmental and social performance aspects. With regard to social performance, some aspects are mentioned that can also be found in the GRI guidelines, such as emphasis on health and safety issues at the workplace, employee participation in decision-making processes, provision of training possibilities and the promotion of minorities.

Commitment to consider those aspects (among others listed) and improve their performance gives companies the chance to be represented in the funds. The index shows a better performance on the market as for instance compared to the DAX (Deutscher Aktien Index) (German Share Index) and the S&P 500 (Standard & Poor's 500), suggesting that responsible behaviour and long-term thinking is economically advantageous.

A second benefit of demonstrating socially (and environmentally) responsible behaviour is with regard to financial institutions such as insurance companies and banks. The business of insurance companies is risk and the lower the risk is, the better they are off. Since CSR implies a better risk management, insurers should welcome those companies, which are already active in this field.

Furthermore, if a company can prove a socially responsible behaviour, the insurer might be more open to lower the premiums than if this was not the case, certainly creating economic benefits for the company in question. With regard to banks operating in a socially responsible way, it might be advantageous as the economic performance of those companies is likely to be better (better risk management, reputation, customer loyalty, etc.), and the risk for banks to invest in them is lower.

Those two examples suggest the close link between the social performance of a company and its long-term economic success.

26.6   Assessing the social performance

In a fast-changing world, it is becoming more and more difficult to determine what exactly is expected from businesses. Due to uncertainty caused by fast changes and the impacts of globalisation, people tend to withdraw their trust from companies. Based on that, a move from a 'trust me' to a 'show me' world can be observed. In cases where scepticisms even grew further, the claim of people towards business is now 'convince me' if there is trust to be established again. One step for the company to meet those expectations is to convince people of their responsible behaviour.

This not only includes proving economic viability and environmental friendliness (e.g., the approach of pollution prevention) but also social correctness and responsible behaviour towards society. If the attempt to behave socially responsibly is to be translated into reality it must be based on some aspects that describe the social performance of an enterprise. Only by this means it is possible to assess and benchmark the situation and achieve continuous improvement.

For general use, the GRI has identified a set of categories and aspects (as well as indicators) concerning the social performance of a company. This classification was done in a multi-stakeholder process and the result is aimed at giving companies guidelines for their sustainability reporting activities. Those guidelines are a response to the challenges of globalisation, which are, for instance, expressed in changing governance structures, since the capacity of existing institutions to manage corporate activity is limited. Social considerations are broken down into the four categories: labour practices and decent work, human rights, society and product responsibility.

26.7 Biotechnology and society

Research in and the application of biotechnology is recognised to have strong societal impacts. A variety of emotions, little interest in the technology itself and a lack of confidence are associated with biotechnology. Nevertheless, there are several social aspects, which are of significant importance to companies operating in the biotechnology sector.

This section presents those aspects, which have been identified based on the theoretical framework of the GRI social aspects and characteristics of the biotechnology industry supported by expert interviews.

26.7.1 Importance of social responsibility for biotechnology

Despite the fact that biotechnology is widely recognised as having great potential for contributing to a more sustainable future, concerns about its application remain. The effects technologies being developed today can have on society could reach dimensions going far beyond our imagination. Consequently, companies performing research and development need to investigate reactions of society to these technologies even before bringing them on the market.

There are risks and uncertainties connected with the employment of living organisms with regard to their ability to change and actively interact with their environment, causing a reserved reaction of the public towards biotechnology. The doubts people have are created partly due to a lack of knowledge and awareness of how the technology works, but are also based on ethical considerations and potential risks associated with this new technology. The main concerns the public has towards biotechnological applications regard plant and animal breeding whereas on the other hand applications in medicine are widely accepted. For the reason that biotechnological innovations and developments might not be acceptable to society, companies need to know society's values, fears and expectations. Having this knowledge can also contribute to saving investments in money, time and human resources as developments of risk-loaded and potentially non-acceptable products/services can be avoided.

On the other hand, if industry can find out and prove that the fears of the public are not based on real risks, companies or industry associations could launch information campaigns to inform people and change their perceptions. As already described (Social Responsibility and Economic Benefits) there are definite economic benefits for companies achievable, especially with regard to the fact that biotechnology companies are interacting with insurance companies as well as banks (e.g., insurance for accidents as they work with living organisms and banks because of the investment need of start-up enterprises).

The vast potential of biotechnology to drive economies towards a more sustainable development has also been recognised by the Commission of the European Communities. Nevertheless, use of this potential can only be made if public support and acceptance is existing and secured for the future. To fulfil this condition, social responsibility demonstrated and lived by companies is essential besides education and information, dissemination by other bodies such as governments, authorities and NGOs. The International Confederation of Free Trade Unions (ICFTU), being concerned with social impacts of industrial operations, shows interest in the biotechnology sector and requires the industry to demonstrate accountability and transparency. One of their requests to the World Summit on Sustainable Development was the establishment of a plan for social and employment impacts of biotechnological developments. Companies can take their share to evaluate and address those impacts and therefore, demonstrate their commitment to social responsibility.

26.7.2 Relevant social performance aspects for the biotechnology industry

The development of the GRI aspects has happened in a multi-stakeholder process and they are meant to be generally useful to all industries. Hence, all of them are likely to apply to the biotechnology sector too. Nevertheless, because of this very general focus of the original GRI aspects some of them may be more relevant with special respect to the biotechnology industry. The relevance of the social aspects for the biotechnology sector in Europe were determined and the most important ones chosen are summarised here again.

Thus, biotechnology:

1. By the use of living organisms has a considerable possibility to contribute to sustainable development and increase our quality of life: This point is unique because only biotechnology uses living organisms and therefore can explore the potential of them.

2. Has big potential for application in cleaner production (through dematerialisation and substitution), but is also applicable for End-of-life purposes: This point is unique in connection with the first point; the fact that living organisms are used.

3. Is knowledge based, requires highly-skilled labour: This point is not unique as those features also apply, for instance, for the ICT sector.

4. Is new, is developing and growing rapidly: This point is not unique because other new technologies do so too.

5. Uses living organisms, therefore shows a (maybe uncontrollable) dynamic, because living things can change (randomly or if outside conditions change), which leads to a uncertainty regarding future impacts: This point is unique as it reflects the definition of biotechnology.

6. Due to the fact mentioned in point 5, it raises concerns among the public but also among scientists (they are scared of the technology and its potential impacts): Unique, because the impacts can have a dynamic character and many of them are not known yet.
7. Requires interdisciplinary teams to explore its potentials: Important but not unique; interdisciplinary teams are increasingly required in many other fields.
8. Raises, to a certain extent, ethical questions, i.e., application of genetic engineering: This point is not unique; ethical questions are also raised, for instance, with respect to the production and use of weapons.

Out of the characteristics of biotechnology some unique features can be identified. Those features are reflected in the points one, two, five and six. Since points one and two together have positive and points five and six have negative implications, it is possible to form two groups out of the four points. One group stands for the large potentials of biotechnology to contribute to sustainable development and thus perhaps to improve the quality of our life (1 and 2). To investigate the relevance of social aspects two criteria can be created out of this:

1. Risk and uncertainty criterion: Concerns and uncertainty about potential (future) impacts (because of the dynamic nature of living organisms and resulting possible changes).
2. Potential benefits criterion: Contribution to sustainable development and improvement of our quality of life.

The reasoning to investigate the specific relevance of aspects to the biotechnology industry is: If certain social performance aspects are specifically relevant for the biotechnology industry, the aspects should address those unique characteristics, i.e., the criteria. The aspect can either be important or relevant with respect to the criterion or might have a low meaning so far. Additionally aspects with uncertain relevance according to available information were identified. The choices will result in a set of social performance aspects specifically important to be addressed by biotechnology companies, aiming at a socially responsible performance.

Bioethics and biosafety

27.1   Introduction

Biotechnology is at the intersection of science and ethics. Technological developments are shaped by an ethical vision, which in turn is shaped by available technology. Much in biotechnology can be celebrated for how it benefits humanity. But technology can have a darker side. Biotechnology can produce unanticipated consequences that cause harm or dehumanise people. The ethical implications of proposed developments must be carefully examined. The ethical assessment of new technologies, including biotechnology, requires a different approach to ethics. Changes are necessary because new technology can have a more profound impact on the world; because of limitations with a rights-based approach to ethics; because of the importance and difficulty of predicting consequences; and because biotechnology now manipulates humans themselves. The ethical questions raised by biotechnology are of a very different nature. Given the potential to profoundly change the future course of humanity, such questions require careful consideration. Rather than focusing on rights and freedoms, wisdom is needed to articulate our responsibilities towards nature and others, including future generations. The power and potential of biotechnology demands caution to ensure ethical progress.

Biotechnology, at its core, is about understanding life and using this knowledge to benefit people. Many see biotechnology as a significant force in improving the quality of people's lives in the 21st century. Obviously, biotechnology is intimately tied to science and scientific knowledge. At the very least, biotechnology promotes a certain vision of life, one in which some things are viewed as good and to be encouraged or pursued, and other things are bad and should be avoided or eliminated. That vision influences people's choices and what is viewed as ethically appropriate. A two-way flow exists in which ethics influences biotechnology even while the science impacts ethics.

At times, the relationship between biotechnology and ethics is portrayed as one of conflict. Sometimes the impression is conveyed that ethics is needed only when someone wants to tell others that what they are doing is wrong. To a degree, this is understandable since controversy, debate and argument are usually integral to ethics discussions.

But ethics is just as important when there is consensus that a direction is good and right. The role of ethics is often invisible at this stage. There was not an ethical debate over whether to search for a cure for cancer. But the decision to pursue such research was motivated by a common vision that curing cancer was the ethical thing to do. Ethical examination of issues is important not only as a form of critique but also to identify and celebrate the right things people do. The effort, resources and creativity focused on developing better treatments are ethically laudable. As such, there is much to celebrate about biotechnology. Society and individuals have benefited in many ways from technology. Many technological developments protect people from illnesses and natural disasters, giving some people 'liberation from the tyranny of nature'. In some parts of the world, people have higher living standards. Travel and communication have developed in unprecedented ways. Many of these changes can be welcomed as ethical developments.

Yet at the same time, other ethical considerations must be considered. At what price are some of these developments realised? Some developments seem motivated by a desire to find treatment at any price. Assisted human reproduction is a particularly controversial area where biotechnological treatment of infertility leads to many ethical dilemmas. Even with less controversial conditions like heart disease or cancer, developments have left people with high expectations that cures should exist. Some are concerned that technological developments lead to dehumanisation or in healthcare lead to less emphasis on caring. Ethical concerns exist about justice, and how fairly these technological benefits are distributed—both within society and around the world. With all the options now available for some, concerns are raised about whether too much choice is bad for us. Overall, though, technology has a strong ethical foundation.

The appropriate response to misgivings and concerns is not to reject technology. 'By turning our backs on technological change, we would be expressing our satisfaction with current levels of hunger, disease, and privation. We simply cannot stop while there are masses to feed and diseases to conquer, seas to explore and heavens to survey'. The benefits of technology, realised and potential, point to a technological mandate: 'Biotechnology should strive to benefit people's lives.' Many of the concerns about technology can be traced to the technological imperative: the idea that something should be developed because we can or we think we can. The distinction between a technological mandate and the technological imperative rests on the ultimate goals of biotechnology. Before addressing whether it can be done, research must answer, 'Why should it be done?'

27.2   Goals of biotechnology

Ethics includes assessment of the rights and wrongs of specific technologies and applications (like cloning or genetic diagnosis). Another important pursuit within ethics is examining the broader goals and aims of enterprises like biotechnology. The relief of sickness is one goal, but there are others that can be more ethically controversial. Aubrey de Grey at Cambridge University has suggested that biotechnology should be directed towards 'engineered negligible senescence'. He stated, 'I am about indefinite extension of longevity Average life-span would be in the region of 1000 years seriously'. De Grey claims that over the next 25 years enough progress will be made in biotechnology to allow people to extend their lives long enough to obtain the next set of benefits. In this way, little by little, people will live longer and longer, effectively preventing death.

Developing the necessary biotechnology for engineered negligible senescence assumes that indefinite life extension is good for humanity. Even if accepted as an ethical goal, it would be one goal among many. Would it be the most appropriate goal for biotechnology? This question is especially pertinent given the limited resources available for biotechnology. Resources are also needed for education, to better distribute the healthcare resources already available, and to provide debt relief for poorer nations. How much investment towards the goal of indefinite life-extension would be in keeping with global justice? While people in developed countries can expect to live into their 80s, the average life expectancy at birth in 2003 was still in the 30s in some African countries.

These types of questions require ethical evaluation. Time should be taken to reflect on the broader implications of pursuing biotechnology. For example, the Centre for Responsible Nanotechnology claims that 'much industry can be directly replaced by molecular manufacturing'. The economic fallout from such developments would be immense, leading to significant social changes with the potential for good and harm. These ethical issues need careful examination even before the technological issues are resolved. Taking the time to reflect on these aspects of scientific developments can be difficult, especially with the pace and focus within biotechnology. The pressures of competing for funding, making breakthroughs, securing intellectual property, and obtaining market share all push against calls for caution or time-consuming reflection. Technological development can seem like a motorway, everyone on the fast track to success. Ethics, even when well intentioned, can seem like a diversion or a roadblock that prevents biotechnology reaching its destination, or delays it inexcusably. However, there is a growing realisation that ethics must be a part of the planning process within biotechnology. In many areas of research, ethics does impact the design of scientific experiments. Any research involving human or animal participants will be scrutinised by ethics committees. The methodology must conform with ethical codes and guidelines. An argument can be made that publicly funded research should be conducted in ways that conform with society's values. 'When the nation decides an activity is worth its public money, it declares that the activity is valued, desired, and favoured.'

Therefore it is important to ensure that what is publicly funded is ethically acceptable in society. The goal of relieving suffering is widely accepted, yet it must be balanced against other societal goals. The ethics of proposed biotechnological developments must be scrutinised carefully.

27.2.1 Darker side

Even such a laudable goal as relieving human suffering cannot be taken as condoning any and all biotechnology. Humanity's creativeness and resource fulness have long been recognised and praised. But human activity can have a darker side. The ancient Greek philosopher Sophocles reflected on these two sides of technological development. On the one hand he noted many human accomplishments in transport, agriculture and medicine. But he also pointed to problems with this same inventiveness. The human capacity for good or evil, whether intended or unintended, impacts how people view the ethics of technology. Hans Jonas fled Germany during the Nazi era and eventually taught philosophy in New York. One of his life's projects was to develop an ethics for technology. His approach was based on his conviction that the new technological age raises several ethical challenges that earlier technology did not have to address. 'Modern technology has introduced actions of such novel scale, objects, and consequences that the framework of former ethics can no longer contain them.' Biotechnology is a particularly fitting example of technology with such fundamentally different characteristics that it requires a careful re-examination of how its ethical dimensions are evaluated. Biotechnology 'raises moral questions that are not simply difficult in the familiar sense but are of an altogether different kind'.

27.2.2 Challenging characteristics of biotechnology

Vulnerability of nature

Jonas contends that ethics prior to the new technological age focused on human–human interactions. Human dealings with the non-human world were regarded as ethically neutral. The capacity for new technology to have global impact shows that ethics needs to broaden its focus. Environmental problems and the existence of nuclear technology demonstrate the importance of ethical examination of more than just human–human interactions. New technology also highlights the vulnerability of nature. Previous technological developments appeared to assume that natural resources were in endless supply and that nature could rebound from any human impact. Environmental changes show these assumptions were problematic. Ethical evaluations of biotechnology need to take the vulnerability of nature into account.

These issues also point to limitations in previous ethical approaches that focused only on humans. At the same time, a concern for these broader issues can lead to new technological challenges and exciting research opportunities, such as has occurred with research into renewal energy sources stemming from ethical concern for the environment.

Limitations with rights

Rights-based approaches to ethics have made important contributions to human welfare. They provide a means by which vulnerable humans can argue for more ethical treatment. However, such approaches have their limitations. A rights-based approach can become very individualistic, with each party focused on his or her rights. Access to biotechnology and new treatments can be defended on the basis of individual rights and personal autonomy. Yet this approach does not lend itself easily to concerns about people seeking treatments that are ethically questionable or of uncertain benefit. For example, individuals may want reproductive cloning, but the concerns of future generations and society as a whole need to be considered. Rights-based approaches are problematic in these situations since rights are typically held by individuals and are not given to those who do not as yet exist. A rights-based approach to ethics must include some method of identifying those who bear rights. Those who have rights place duties on others to uphold those rights. It has proved very difficult to find consensus on how rights are to be ascribed. One approach is that all humans are inherently entitled to all human rights. This raises questions about when a human is given these rights (at fertilisation or birth or some other point).

It also leaves no guidance on how to treat the non-human world. Biotechnology requires answers to these questions to address ethical concerns about non-human species and nature as a whole. This has led to an approach where rights are granted based on particular abilities and attributes. There is little consensus over what abilities entitle an organism to rights. Philosophically, it is also difficult to justify why any particular attribute should lead to the granting of rights. The whole approach is criticised as being motivated by a desire to treat unethically those not given rights. This is particularly relevant to research on human embryos, especially embryonic stem cell research.

Developments in biotechnology point to serious limitations with a rights-based approach to ethics. Rather than providing insurmountable problems for ethics, these point to the need for a different approach to ethics. Jonas and others point out that rather than focusing exclusively on human rights and entitlements, the new technological era requires a greater focus on human responsibility.

Future consequences

Earlier technology impacted humans and their lives, but did not have the potential to change human nature. Biotechnology does. With that comes the potential for broader and long-range consequences. Predictions about these consequences can be difficult and unreliable. This is particularly cogent with genetic technology. The consequences of our ability to manipulate the human genome could impact many, if not all, future generations. The way genes interact with one another means that manipulating one gene could have unintended effects on other genes or their expressed proteins. This is especially important given the recent realisation that the human genome contains fewer genes than originally presumed. Biotechnologys mistakes may produce problems, but so too might its successes. As technology has developed and spread, 'the more all of reality is seen as matter-of-factly material and hence as controllable in a completely technical and rational manner'.

Successful technological solutions could lead people to view all our problems as needing a technological fix. The medicalisation of patients and the instrumentalisation of people are consequences of technologys successes. This can have a dehumanising effect on human life, which makes it easier to treat some humans as less than fully human. This is a way in which technology can take on a life of its own and have much more profound ethical consequences.

Biotechnology has the added capacity to produce products that literally do take on life. The technology humans developed in the past was inanimate and could be left unused if found to be ethically problematic—as difficult as that might have been. However, biotechnology now makes possible the creation of products that are themselves alive. 'The work of [human] hands takes on a life of its own and independent force, no longer figuratively but literally'. The living products of biotechnology are no longer under human control in the way an inanimate machine was. Now the living product itself could influence its impact and might develop into new forms of life with unexpected consequences (although such problems have not developed to date with genetically modified bacteria). Such factors should remind us of the place of awe and mystery in the face of nature. We humans are limited in our ability to understand, control and direct nature. That realisation should cause us to pause before attempting to manipulate life through biotechnology. It should lead to a sense of caution. Yet often the very opposite is the case, with the pressure to rush to be the first to develop something new. The precautionary principle is particularly pertinent with experimentation on humans.

Impact on human nature and personhood

No area of biotechnology more clearly brings to focus the need for careful ethical reflection than its potential to impact human nature. Previous technology has provided new tools that impacted human activities and society. Humans were the makers of technology. Some aspects of biotechnology now make humans the objects of technology. Humans have turned upon themselves and are ready 'to make over the maker of all the rest'. The capacity for biotechnology to create and change human lives calls for careful reflection on what it means to be human and the place of human personhood.

Recent developments with stem cell research and cloning have been the lightning rod for debate over human personhood. These discussions point to the gulf between proponents on the different sides. Some have viewed embryos as 'featureless bundles of cells'. From this perspective the human embryo is a human non-person that can be used and destroyed in research. Others disagree and maintain that the human embryo should be treated as a person, making it unethical to treat it merely as a means to others' ends. Personhood can be viewed as an inherent attribute of all humans. This confers all humans with certain rights and determines how persons should be treated ethically. This approach protects humans, especially the vulnerable, from unethical treatment. The other approach makes personhood conditional on reaching some stage of development or possessing certain abilities. Only humans with those capacities are then entitled to protection.

A fundamental problem with this approach is that it always arises to justify killing those declared to be human nonpersons. How will it affect us to treat human lives as commodities to be manipulated and destroyed at will? When we justify doing so with embryos, will it become easier to do so at later stages of development?

This debate points to the difficulty of determining public policy when sections of society have irreconcilable positions on matters of fundamental importance. We must also examine how biotechnology itself impacts our view of human nature. Leon Kass asks how will it affect us 'to look upon nascent human life as a natural resource to be mined, exploited, commodified. The little embryos are merely destroyed, but we—their users—are at risk of corruption'. This is much more than a debate over rights. This is about human dignity, including what it means for humans to act with dignity. This changes the focus from ascribing rights to determining responsibilities.

27.2.3 Central place of responsibility

The enormity of the potential impact of biotechnology on human nature should cause us to proceed cautiously. Biotechnology has the potential to do great good. But it also has the potential to cause much harm. This could arise in the physical realm through unexpected consequences of the technology itself. But other harms could arise through the non-physical impacts of biotechnology. Cars and computers have affected many aspects of human life and society.

Biotechnology could change what it means to be human. A rights approach to ethics makes clear where people have rights. Each right carries a corollary duty or responsibility. If people have a right to healthcare, someone has the responsibility to provide healthcare resources. Much energy has been expended identifying and defending human rights. We now need a similar emphasis on human responsibilities.

Responsibility is also a corollary of power. Biotechnology brings new powers to humanity. These powers should remind us of our responsibility to nature and the environment, to all of life, to the future, and to human nature and personhood. To understand these responsibilities entails the development of wisdom. That wisdom requires ethical reflection before developing specific forms of biotechnology. Taking the time for that reflection can go against the pace of biotechnological developments and hubris over human wisdom.

Jonas warned that new technology was propelling us towards a utopian future. Aubrey de Grey exemplifies that vision for biotechnology. These developments have the potential for much good, but also risk changing, harming or even destroying some species, including ourselves. To make the right ethical decisions 'requires supreme wisdom—an impossible situation for man in general, because he does not possess that wisdom, and in particular for contemporary man, because he denies the very existence of its object, objective value and truth. We need wisdom most when we believe in it least.'

Jonas was referring to the post-modern rejection of objective truth that has become so prevalent—the idea that all answers are equally valid. In contrast, ethics searches for better answers to ethical questions. It acknowledges the limitations in current wisdom, and strives to improve our understanding. The way forward is muddied by our inability to accurately predict the consequences of proposed biotechnological developments. Some argue that we should push ahead and deal with problems as they arise. But given the scale of disaster that biotechnological mistakes could trigger, Jonas' guiding principle contains much wisdom. He argued that 'ignorance of the ultimate implications becomes itself a reason for responsible restraint—as the second best to the possession of wisdom itself'.

Time and resources must be committed to examining the ethical implications of proposed biotechnological developments. The potential impact on all aspects of nature must be considered. The social, emotional and spiritual implications of developments in biotechnology must also be examined. When humans themselves are the objects of biotechnology, great caution is necessary lest we promote a view of ourselves and our neighbours as nothing more than living bits of technology.

27.2.4   Ethical and moral issues

Intellectual property protection is one of the social systems that has evolved in modern society. Like the technology that it is applied to protect, it is a system that needs to be subject to ethical analysis to examine whether it is suitable for a moral society. The principle benefit claimed for patents is that rewarding an inventor creates a positive environment for progress of research that leads to the betterment of society. If this is true than this is consistent with the ethical principle of beneficence. History suggests that the financial interest in a free market creates more funding for research, and faster overall progress in research in important areas has been the result of the intense research efforts. This point has been used by industry to oppose moves to block patents on biotechnological inventions that arise from other ethical concerns. The issue is however more complex than a simple examination of the benefits of intellectual property to one society, because there are always winners and losers in trade. We have to consider the ethical principles of justice, and non-maleficence. Even more complex is deciding just who are the actors involved in the equation.

Some key ethical issues in patenting in scientific research include:

1. Is the principle of beneficence or loving good, served more by having research than by not having research?

2. Do we encourage more research into more beneficial areas of science by the incentive system of patents than we would by not having patents?

3. Is justice served by systems of intellectual property protection?

4. How can we justly reward all the inventors in the often long process of developing a useful product? Should we only reward the final step, and how to value farmer's innovations in the development of plant and animal varieties?

5. How to value indigenous knowledge, and to share the benefits with the communities whose ideas gave raise to pursuit of a new product, for example with medicinal plants?

6. What are the tolerable limits of doing harm by research subject, e.g., animals including humans?

7. What are the tolerable limits of doing harm by rigid enforcement of patents if price becomes a barrier to use of a product by persons in need?

8. Ethically can anyone own a product of their mind, a product of nature, a product of a designed process, a discovery or even an invention?

9. Does it make any difference whether the product or process involves living organisms or rocks?

10. Should we expect the practical law to share the same goals as that of ethics, namely can we expect ideal ethical laws or some compromise?

27.2.5 Moral arguments supporting patenting

In this section we will explore some of the arguments that go beyond beneficence. The debate over the patenting has been intense, but it remains unsettled dispute the positions taken. We expect the debate to continue even though public policy has demanded that legal measures be adopted to guide the application of patent law to biotechnology.

One of the ethical arguments expressed when supporting patenting of biotechnology inventions is that patent law regulates inventiveness, not commercial uses of inventions. Those arguing this position would claim that we need to encourage an inventive society so that we can progress knowledge into the future. Rewarding inventions means some members of society can devote their time and energy to creativity. The commercial use of inventions, is not dependent upon the patent itself, that is subject to social demands.

The encouragement of inventiveness, through patenting, means that it promises useful consequences (e.g., new products/research). To develop new tools is a good for society, although not all new products are necessarily good. However, if people and society are really given a choice on which products to use is decided by commercial factors, and could be questioned. The ideal of free market choice is too simple to explain the real world. Support for patents is often seen when national economy is defended. If other countries support patents, then our country needs to also, if the biotechnology industry is to compete in a global market. In a following section we will discuss the morality exclusion in patent laws. There is a balance between an ideal of sharing all knowledge with each other and a giving society versus the harsh reality that countries are engaged in tough international competition that a modern free market has created. It is almost impossible for any country to stand on its own in a global system, and the institution of the World Trade Organisation (WTO) and the Agreement on Trade Related Aspects of Intellectual Property Rights (TRIPS) have made this legally binding.

One argument that is often neglected by opponents of patenting is that if patenting is not permitted, useful information will become trade secrets. Patenting actually forces an invention to become public knowledge, so that other researchers can begin to investigate knowledge from the new invention. If a company believes it can keep an invention secret for more than twenty years then it may be better economic strategy to keep that knowledge a trade secret. The indirect results will be affected by factors such as whether during the period of patent exclusion certain companies have been well established and are able to provide beneficial services or monopolies, and the relative advantages of the open knowledge after patenting compared to industrial secrecy that could occur if patenting was difficult to obtain.

More significantly, there may be a greater amount of total knowledge and a more rapid completion date using the approach involving private companies. The means by which these approaches are pursued differs. In one approach the government laboratories spend their resources on particular focused projects like the genome project at the expense of other projects, but with the cumulative results being openly available to all. In the other approach, the private companies do the research, which would create more total biomedical research knowledge, but certain parts of this would be tied up in patents, though the knowledge would also be available with a small delay.

We have seen the competition between genomics companies like Celera and the international government genome project led to the completion of the genome sequence several years ahead of schedule, in the year 2000, instead of the planned 2005 completion date. While there were gaps in the initial genome sequence that resulted from the race, it still allowed completion several years in advance. Another example is competition between genomics companies and a consortium of major pharmaceutical companies in the development of maps of single nucleotide polymorphisms (SNPs). It was novel to see competitors in pharmaceuticals funding together the SNP consortium so that the SNP maps would be public domain. The motive is commercial, because having the SNP map public then allows companies to explore individual SNPs for medical treatment without having a complex web of royalties to pay to genomics companies that are attempting to patent as many SNPs that they can before the SNPs they claim are listed in the public domain. Another argument follows the ethical principle of autonomy and is that patenting rewards innovation. It says that if I invest something I have some right to use that knowledge first. It is ethically weak in philosophy, however, it is consistent with the working philosophy of individualistic modern society where people are rewarded for their hard work. The harder we work the more money we are said to be able to claim, the spirit of entrepreneurship. It is the common morality of Western society where individual work is rewarded.

27.2.6   Ethical arguments against patenting

Despite the arguments for patenting above, this issue remains contentious and the fact that different countries have conflicting policy reflects this. The issue is closely related to the commercialisation of biotechnology, but some sort of information protection is accepted as an incentive to invest in research of benefit to society. The arguments against patenting include a variety of arguments in response to the questions raised. Important to many are metaphysical concerns about promoting a materialistic conception of life through economic valuation. However, farming has traded animals and medicine has traded medicines, for millennia. Until the convention on biological diversity made local living organisms the intellectual property of the states, there was the concept that living organisms are the common heritage of humankind, so they should not be exploited. Opponents of patenting claim it promotes inappropriate human control over information that is common heritage.

Patenting is said to produces excessive burdens on medicine. These include increased costs to consumers, and payment of royalties for succeeding generations. The claim that the function of patents is to regulate inventiveness rather than to regulate commercial uses of inventions is perhaps avoiding the consequences of the system. There have been some controversies regarding the commercial monopoly held by the company which was able to patent AZT, the initial HIV/AIDS treatment, which gained large profits in view of its monopoly. It is all the more questionable whether this should be allowed because of the key roles that government funded research played in developing AZT and showing it was active against AIDS. There are other examples where the commercial monopolies obtained cannot be said to be in the best public good, and the existence of patent laws is certainly relevant to the later commercial uses of inventions. The economic system is tied to patent recognition, and we need to consider the broad social consequences.

In fact a number of these above issues may not be directly affected by permitting patents, as the issues like the distribution of wealth and international competitiveness exist independent of the patent debate. A general criticism of modern biotechnology is that it may encourage technology to be done for its own sake, and for the creation of new markets and employment, rather than seeking the best solution to a problem in terms of the environment or human society. The assumption that wealthier societies protect the environment more than poorer ones is flawed, especially when we see the rising consumption patterns of richer societies. It is not really clear whether the type of research supported by patenting and the research investment it protects, is the research of most benefit to society. Medical and agricultural products are clearly needed, but not always the most efficient and sustainable processes and products are used, as seen for example in chemical pesticide industry or expensive pharmaceutical alternatives to existing medicines. We can also consider the amount of money people in developed countries spend on luxury products, such as cosmetics, that may be considered a waste of research investment in terms of distributive justice, when compared to life threatening diseases. Is the sequencing of DNA really an invention? The DNA could be viewed as a random sequence of bases, and the author is the sequencer, but this is not what we would normally talk of as an author or inventor, rather the sequencers are discoverers. In the days of colonial rule a discoverer could claim a land as their property, but later it was recognised that the pre-existing people had claims to the property no matter how it was developed by the colonisers. The sequencers of DNA are not sequencing unowned land but rather they are sequencing uncharacterised land, the name of mappers is rather suitable for this analogy. Some critics of ownership could go as far as to call those who seek to profit and to control the decisions concerning the human genome project without general consultation, a type of 'genomic imperialist'. The DNA is not random, it is merely unknown. This is an important difference, in addition to the common possession of the DNA sequence by every member of humanity, the sequencers are not authors. While it may also be unconventional to call the possessor of information the author, they have more claims to that title than the sequencers, in addition they can be called the owners (only in this general way the human sequencers are shared owners because they also possess DNA).

27.3   Biosafety

Biosafety is prevention of large-scale loss of biological integrity, focusing both on ecology and human health. The international Biosafety Protocol deals primarily with the agricultural definition but many advocacy groups seek to expand it to include post-genetic threats: new molecules, artificial life forms, and even robots which may compete directly in the natural food chain. Biosafety in agriculture, chemistry, medicine, exobiology and beyond will likely require application of the precautionary principle, and a new definition focused on the biological nature of the threatened organism rather than the nature of the threat. When biological warfare or new, currently hypothetical, threats (i.e., robots, new artificial bacteria) are considered, biosafety precautions are generally not sufficient. The new field of biosecurity addresses these complex threats. Biosafety level refers to the stringency of biocontainment precautions deemed necessary by the Centres for Disease Control and Prevention (CDC) for laboratory work with infectious materials.

Objectives of safety guidelines: The National Institute of Health (NIH, USA) has developed guidelines for recombinant DNA research with a view to specify the practices for constructing and handling:

1. Recombinant DNA molecules.

2. Organisms and viruses containing recombinant DNA molecules. These guidelines require approval and clearance of any recombinant DNA experiment that requires such approval/clearance from NIH or another Federal agency.

All experiments involving the deliberate transfer of recombinant DNA or DNA or RNA-derived from recombinant DNA into human subjects cannot be initiated without submission of the required information to NIH and other specified agencies. Cartagena Protocol on Biosafety to the Convention on Biological Diversity has been designed to cover the transboundary movement, trails it, handling and use of all living modified organisms that may have adverse effects on the conservation and sustainable use of biological diversity, taking into account risks to human health. The ultimate objective of guidelines that regulate research and development activities using recombinant DNA technology should be to minimise risk from such activities and, at the, same time, encourage these activities.

27.4 Biosafety during industrial production

Many of the processes of manufacture have the potential to generate biohazards. Therefore, it is not surprising that several noninfectious illnesses are associated with biotechnology industrial products/processes. Inhalation appears to be the most significant mode of entry of microbes or microbial products into the body, while entry by ingestion or skin contact can readily be prevented by good operating practice.

Most reported health problems are associated with downstream processes, and the risk of allergic reaction is the greatest after the product is more concentrated. During extraction of intracellular enzymes, the greatest risk is posed by the processes of handling the large quantities of cell debris. An effective monitoring strategy should involve environmental assessment of emissions at all stages of a manufacturing process, and a critical assessment of the biosafety performance of process equipments; in addition, it should maintain adequate surveillance of worker health.

27.5 Risk assessment

Risk is a function of the level of hazard and the probability of occurrence of the hazard. Thus risk may be defined as the undesirable consequences of an activity in relation to the likelihood of these consequences being realised. A hazard is any imaginable adverse effect that can be identified and measured. Risk assessment involves determination of the potential and anticipated adverse effects of the recombinant DNA research to the concerned workers, and of the products of such research on human health and the environment consequent to their accidental or deliberate release (in case of living organisms) or as a result of their consumption. Risk assessment should be carried out in a scientifically sound and transparent manner, and should be in accordance with recognised risk assessment techniques.

27.5.1   Assessment of risk during laboratory research

The chief objective of risk assessment at this stage is to decide on appropriate laboratory procedures, and physical and biological containment levels for the proposed research. The assessment of risk during laboratory research is usually done in two steps:

1. Initial risk assessment.
2. Comprehensive risk assessment.

27.5.2   Initial risk assessment

The initial risk assessment is made by the investigator on the basis of the risk group (RG) to which the organism, on which he proposes to conduct experiments, belongs. Organisms are classified into four risk groups based on their potential effects on a healthy human adult.

1. Risk group 1 (RG1) agents are not associated with disease in healthy humans.
2. Organisms included in RG2 are associated with such human diseases, which are rarely serious and for which preventive or therapeutic interventions are often available.
3. RG3 agents are associated with serious or lethal human diseases for which preventive or therapeutic interventions may be available.
4. Organisms belonging to RG4 are likely to cause serious or lethal human disease for which preventive or therapeutic interventions are not usually available.

27.5.3   Comprehensive risk assessment

Following the initial risk group based risk assessment, a comprehensive risk assessment should be done. This assessment takes into account the following:

1. Experimental organism factors: These factors include such features of the organism used for the experiment as virulence, pathogenicity, infectious dose, etc.
2. Type of manipulation: Careful consideration should be given to the types of manipulations that are planned for some higher RG agents.

27.5.4   Risk assessment for planned introduction

Risk assessment for planned introduction of living organisms modified by recombinant DNA technology is done to identify and evaluate the potential adverse effects of such introductions on the environment, including biological diversity and human health. Risk assessment should be done on a case by case basis. Risk assessment takes into account the relevant technical and scientific details regarding the characteristics of the following factors:

1. The biological characteristics of the recipient or parental organisms, including taxonomic status, centres of origin and of genetic diversity, habitat where they are likely to persist or proliferate, etc.
2. Taxonomic status and the relevant biological characteristics of the donor organisms.
3. Characteristics of the vector, including its origin, source and host range.
4. Genetic characteristics of the DNA insert the function it specifies and/or characteristics of the modification introduced.
5. Identity of the GMO, and the differences between biological characteristics of the GMO and those of the concerned recipient or parental organisms.
6. Suggested detection and identification methods, and their specificity, sensitivity and reliability.
7. Intended use of GMO, including new or changed use in comparison to the recipient/parental organism.
8. Information on the location, geographical, climatic and ecological characteristics, including relevant information on biological diversity, of the receiving environment, i.e., the environment in which planned introduction is proposed.

In case of transgenic plants, the following are the main concerns related to their planned introduction:

1. The likelihood of a transgenic becoming a weed.
2. Likelihood of the transgene being transferred to wild relatives. Transfer of transgenes conferring herbicide resistance to weeds would be rather hazardous.
3. Introduction of novel genes may cause perturbation of the ecosystem.
4. Virus resistance genes may provide opportunities for evolution of newer virulent strains by recombination, etc.
5. Insect resistant plants would be cultivated on a large scale; this may lead to the evolution of insect biotypes resistant to the toxic effects of the concerned transgene.
6. It is also feared that transgenic plants may affect the flora and fauna of their phyllosphere as well as rhizosphere.
7. The product of transgenes may pose health hazard, e.g., allergy, to human consumers.

Molecular biologists and ecologists represent diagonally opposite views not only on the perception of risk but also about the manner in which risk should be assessed. Molecular biologists view modern biotechnology as an extension of classical genetics, and believe that the risk can be evaluated by studying the features of transgenes and host organisms. But ecologists argue that there exist many uncertainties concerning the risks, and they emphasise the study of entire ecosystem in which biotechnological products, especially recombinant DNA products, are to be released.

27.6 Risk assessment for biotechnology products

In USA, foods and additives developed by biotechnology are subject to the same rigorous safety standards as are conventional foods and additives, and there are no special requirements for such products. The foods and additives obtained from genetically modified organisms are called genetically modified food (GM food).

Assessment of risk from GM food takes into consideration the following:

1. The food crop that has been modified.
2. The potential for any introduced DNA to encode harmful substances.
3. The safety of proteins encoded by *trans* genes.
4. Proof that known plant toxicants and important nutrients are within acceptable levels in the new variety.

There is some chance that proteins encoded by transgenes may cause allergies in some individuals; this is particularly so for proteins derived from such foods as milk, eggs, wheat, fish, tree nuts and legumes. There are some instances where a transgenic variety had to be withdrawn due to severe allergic reactions in some people, e.g., a soyabean transgenic variety in Mexico. Information on the following four aspects of the production process and the product would help in evaluation of the safety of a product generated by recombinant DNA technology.

1. Genetic backgrounds of the host and the donor organisms.
2. The technical procedures used for developing the products.
3. Composition of the food, including any potential toxicants and significant nutrients.
4. Relevant toxicological data on the host organism.

Japan has decided to 'label' those GM foods that are found to be 'safe'. Similarly, European Union has adopted legislation requiring mandatory labelling of all GM food. In contrast, US Food and Drug Administration (FDA) requires the food to be labelled as genetically modified only if the changes introduced through genetic engineering have an impact on the safety or nutrition of the food itself. Labelling is considered desirable because it gives the consumer the choice of deciding if he wishes to have GM food, but companies fear that labelling may frighten the consumers.

27.7 Biosafety containments

Biological containments specifically aims at making such genetic changes in GMOs that reduce the hazard from these organisms when they are accidentally released from laboratories or accidentally transported by workers. Initially, NIH guidelines required the use of *E. coli* strains and vectors that were severely debilitated, viz. that could not infect humans, did not survive and spread outside the laboratory, and could not transfer the introduced foreign genes readily into other organisms in the environment.

These objectives can be achieved by using the following:

1. Auxotrophic mutants of *E. coli* (limits survival following escape of bacteria).
2. *recA* strains (they lack recombination function).
3. Plasmid vectors that are nonself-transmissible and non-mobilisable (such vectors cannot be transmitted to other bacteria).
4. *E. coli* strains not carrying a transposon or antibiotic resistance gene.

Thus biological containment is based on the vector (plasmid, organelle or virus) used for the construction of recombinant DNA, and the host (bacterial, plant or animal cell) in which the vector is propagated in the laboratory; therefore it is also called HV system (host-vector system). Any combination of vector and host, which is to provide biological containment shall be chosen or constructed so that the following types of 'scapes' are minimised:

1. Survival of the vector in its host outside the laboratory.
2. Transmission of the vector from the propagation host to other non-laboratory hosts.

27.7.1 Biosafety physical containments

Physical containment consists of the use of special laboratory design, containment equipments and special operational procedures to restrict the number of organisms accidentally released during normal laboratory operations and to prevent contamination or infection of laboratory workers. Physical containment is grouped into four categories, viz. BL1 (Biosafety Level 1), BL2, BL3 and BL4, respectively. BL1 is applicable to while BL4 represents the most stringent containment levnonpathogenic organisms. As the biosafety level increases, the laboratory design and facilities, and the operational procedures become more and more rigid and stringent so as to minimise the risk of accidental release of the GMOs as well as infection of laboratory workers. The guidelines provide a detailed description of the laboratory design, equipments, etc. for the various biosafety levels. Emphasis is placed on primary physical containment, which is provided by laboratory practices and containment equipment. Special laboratory design provides secondary physical containment for protection against accidental release of organisms into the environment; it is used where experiments of moderate to high potential hazard are performed.

Several different combinations of laboratory practices, containment equipment and special laboratory design may be appropriate for containment of specific research activities.

27.8   Planned introduction of genetically modified organisms (GMOs)

Planned release of GMOs in the environment may be viewed in two different ways. In a process based approach, all GMOs, are considered to pose potential hazards, some of which may be unforeseeable, to the environment. But in product based approach, all GMOs are not recognised as a special category because they present no environmental hazards beyond those already identified for existing products. Biotechnology companies favour a product based regulation for obvious reasons.

27.8.1   Field trials with genetically modified plants

In case of plants conventional breeding cell and tissue culture and genetic engineering share some major similarities in the method of generation the magnitude and type and the methods of detection of the genetic changes but they do differ in some important features.

Usually products of both tissue culture and genetic engineering are subjected to few to many back crosses and selection and before that they both are subjected to tissue culture. The risk of genetic engineering products lies in the possible unexpected effects of the exotic genes in the new cell environment and the new ecological surroundings.

The type of regulation necessary during field trials should depend on the ability of modified plant to survive disperse reproduce and hybridise with crops and wild plants and on our ability to confine the plants to the test site. In case of cross pollinated species, isolation by distance is rarely sufficient to prevent interpollination. In case of viral coat protein gene-mediated virus resistance fears of another virus acquiring the engineered protein coat and thereby produce hybrid viruses is a possibility. Transgenic varieties of crops expressing *cry* genes for insect resistance are in commercial cultivation. *Cry* proteins are rapidly degraded by the stomach juices of vertebrates, but they could have harmful effects on nontarget insect species, e.g., honeybees. Oilseed rape (*B. napus*) expressing C_pTI (Cowpea trypsin inhibitor), chitinase and β-1,3-glucanase has been used for assaying the impack of these genes on honey bees. Chitinase did not affect learning performance, while C_pTI and β-1,3-glucanase had detrimental effects.

27.8.2 Planned introduction of genetically engineered micro-organisms (GEMs)

There is some data available from field research using GEMs. The GEMs that are used live in the environment include bioremediation agents that degrade chemical waste. Immunoflurescence techniques have revealed a much higher survival of microbes than was previously suggested by the conventional techniques. Therefore, it has been advocated that GEMs should be constructed that contain active biological containment systems that minimise their survival and dissemination and/or prevent the dissemination of recombinant DNA outside the target area. An example of such a biologically contained GEM is provided by *Pseudomonas putida* engineered to degrade alkyl benzoates.

This strain contains the E. coli gene gef: Gef protein causes a collapse of the membrane potential, which leads to cell death. The *gef* containment system consists of the following elements:

1. The *xylS* regulator gene, which is expressed constitutively.
2. The regulator gene lad linked to the promoter *Pm*.
3. The structural gene gef driven by *E. coli* promoter *Plac*.

Gene xylS is expressed constitutively: It encodes an inactive regulator protein that becomes active on interaction with *m*-methylbenzoate (3-MB). In the presence of 3 MB, therefore, the *xylS* regulator is active, and it activates transcription of gene lad driven by *Pm* promoter.

The *lacI* repressor, in turn, binds promoter *Plac* and prevents transcription of the *gef* gene linked to it. Therefore, in the presence of 3-MB, transcription of the *gef* gene is repressed; as a result, *gef* protein is not produced, and the *P. putida* cells survive normally as long as 3 MB is available to them.

However, when 3 MB is not available to a cell, the *xylS* activator protein remains inactive. As a result, the regulator gene lad is not transcribed and lad repressor is not produced.

In the absence of lad repressor, *gef* gene (driven by *Plac* promoter) is expressed, and *gef* protein is produced. Gef causes collapse of the membrane potential, which leads to cell death. Therefore, the cells of this genetically engineered *P. putida* strain commit suicide as soon as 3 MB is exhausted and becomes unavailable to them; this minimises their survival in the environment after they have degraded the alkylbenzoates present in the environment.

Similarly, an appropriately regulated endonuclease gene can be used to limit GEM survival and also to prevent transformation by GEM DNA after cell death. The first release of a recombinant micro-organism was done in 1986, it related to a Tn5-containing strain of *Pseudomonas* fluorescens in the Netherlands. Since then more than two dozen planned introductions have been done. The initial results from these studies suggest that planned introduction of recombinant bacteria can be done in a safe and responsible manner.

27.9 Biosafety guidelines

In, the ministry of environment and forests (MOEF) promulgated in December, 1989 the rules and procedures for the manufacture, import, use research and GMOs as well as products made from such organisms, this was done under the provision of the environment protection Act, 1986 (EPA).

Anti-violation and nocompliance, including nonreporting of activities, in this area would attract the punitive actions provided under the EPA. In addition, the Indian recombinant DNA safet guidelines and regulation have been prepared by and are available on request from recombinant advisory committee (RDAC), Department of Biotechnology (DBT), New Delhi. Field trials using transgenic plants began in 1995. The DBT implements the research and development experiments utilising GMOs and recombinant DNA products, while MOEF implements the large scale commercial use of these, the silent features of these guidelines and Regulations are as follows:

1. Every organisation involved in research and development using recombinant DNA technology is required to set up an institutional biosafety committee (IBC), which has a DBT nominee, IBC is the nodal point for interaction of the organisation with the government.

2. The DBT has a review committee for genetic manipulation (RCGM), which has all the approvals of ongoing projects on GMOs and several other issues related to recombinant DNA research and development.

3. Each state has, in addition, a state biotechnology coordination committee, and a district level committee, which are involved in inspection and monitoring of experiments at the field sites.

4. The MOEF has an interministerial committee called the genetic engineering approval committee (GEAC), which has subject specialist as members, and is the competent authority to decide on the large scale use of GMOs.

5. The guidelines recognise four levels of risk in the case of experiments with micro-organisms, based on pathogenicity of the micro-organisms, local prevalence of the concerned disease and of epidemic causing strains in India.

6. Experiments with micro-organisms, plants and animals are grouped into the following categories:

   (a) Exempt category (for self-cloning experiments).

   (b) Category requiring intimation on initiation to competent authority (experiments involving nonpathogenic DNA vector systems).

   (c) Category requiring review and approval by the competent authority (cloning of genes for toxins, antibiotic resistance, etc.).

7. Four different biosafety levels are recognised and containment facilities for each level are recommended for necessary safeguards.

8. Physical containment envisages to limits the spread of dangerous micro-organisms by:

   (a) Good laboratory practices.

   (b) Safety equipment.

   (c) Laboratory design and facilities.

9. Biological containment consists of the use of vector and hosts in such a way so that it:

   (a) Can limit the infectivity of vector to specific hosts.

   (b) Control host vector survival in the environment.

Biohazards from biotechnology industries are given in Table 27.1.

Table 27.1: Biohazards from biotechnology industries.

Operation	*Hazard*
Raw material weighing, mixing, etc.	Allergic dusts or aerosol
Bioreactor fermentation	Aerosols from the reactor, spillage, effluent, contamination, spillage, leakage
Biomass separation by centrifugation, filtration	Aerosol, spillage, leakage
Product purification	Aerosols, spillage
Product blending, filtration, drying, packaging	Aerosols, dust, spillage
Effluent sterilisation, disposal	Discharge of untreated effluent

Noninfectious illnesses associated with production processes biotechnology industries are given in Table 27.2.

Table 27.2: Noninfectious illnesses associated with production processes biotechnology industries.

Product/Process	Noninfective illness
Antibiotic	Cardiovascular disorders, candidiasis, etc.
Brewing	Dermatitis, malt fever
Citric acid	Asthma, bronchitis
Enzymes	Asthma, conjunctivitis, dermatitis
Endotoxin	Flu symptoms
Fungal fermentation	Asthma, bronchitis
Single cell protein	Allergic, asthma, dermatitis
Steroids	Feminisation of males, hypertension, increased body weight

Regulating the use of biotechnology

28.1   Introduction

Intellectual property is the term used to describe the branch of law which protects the application of thoughts, ideas and information which are of commercial value. It thus covers the law relating to patents, copyrights, trademarks, trade secrets and other similar rights. The development of the genetic resources of biodiversity is known as biotechnology. Broadly defined, biotechnology includes any technique that uses living organisms or parts of organisms to make or modify products, to improve plants or animals, or to develop micro-organisms for specific uses. Mankind has used forms of biotechnology since the dawn of civilisation. However, it has been the recent development of new biological techniques (e.g., recombinant DNA, cell fusion and monoclonal antibody technology) which has raised fundamental social and moral questions and created problems in intellectual property rights.

Intellectual property protection for biotechnology is currently in a state of flux. Whilst it used to be the case that living organisms were largely excluded from protection, attitudes are now changing and increasingly biotechnology is receiving some form of protection. These changes have largely taken place in the US and other industrialised countries, but as other countries wish to compete in the new biotechnological markets, they are likely to change their national laws in order to protect and encourage investment in biotechnology.

There is at the moment no clear international consensus on how biotechnology should be treated. Although bodies such as the World Intellectual Property Organisation (WIPO, the United Nations permanent body primarily responsible for international cooperation in intellectual property) and the Organisation for Economic Cooperation and Development (OECD) have conducted separate studies and produced various reports, these have only sought to make governments more aware of the potential problems and to offer some suggested solutions. In view of the highly controversial nature of providing intellectual property protection for biotechnology, it is likely that in the short-term developments will be at a national and regional level.

28.2   Intellectual property protection currently available

There are currently two main systems of protection for biotechnology: Rights in plant varieties and patents.

Both systems provide exclusive, time-limited rights of exploitation and are described in more detail below.

Keeping biotechnology 'secret' can also be a valuable form of protection. National treatment of trade secrets is diverse and all attempts to harmonise trade secret laws in Europe, for example, have failed. Most jurisdictions do provide some form of protection against those who steal or use others trade secrets unfairly. However, the problem with this form of protection is that the secret generally becomes public once the biotechnology is used commercially and thus the protection is lost.

It is conceivable that the law of copyright could afford some protection for biotechnology. Lines of genetic code are analogous to some extent with computer programme code, which has now been incorporated into the copyright systems of most industrialised countries. However, this route to protection is fraught with practical and conceptual difficulties and is generally thought to be unsuitable. There is as yet no recorded case of biotechnologists claiming copyright in their inventions.

Trademarks are also unlikely to be of much use in protecting biotechnology, though they may of course prove important later in regard to marketing products, processes or services. An attempt to register the name of a plant or an animal as a trade mark is unlikely to be successful as public policy would prevent it (in England, registrations for names of varieties of roses have been removed from the Trade Mark Register for lack of distinctiveness and because of the likelihood of confusion).

28.3   Rights in plant varieties

Prior to the mid-1960s only a few countries (e.g., Germany, U.S.) gave any intellectual property protection to plant varieties. Because of pressure from their plant breeding industries, 10 western European countries entered into a diplomatic process in the early-1960s which eventually culminated in the formation of an International Union for the Protection of New Varieties of plants (UPOV) and the signing of a Convention (the UPOV Convention 1961). Since that time a number of other countries have become parties to the UPOV Convention. Amendments were made to the UPOV Convention in 1978, principally to facilitate the entry of the U.S..

The UPOV Convention requires that each member country must adopt national legislation to give at least 24 genera or species protection, in accordance with the provisions of the convention, within eight years of signing. A plant variety is protectable (a protectable variety) under the UPOV system if it is distinct, uniform, stable (DUS) and satisfies a novelty requirement. Novelty and distinctiveness equate broadly to novelty under patent law, but are more leniently applied in comparison to the patent rule. Satisfaction of the DUS criteria is conducted by the national authority responsible, usually by growing the variety over at least two seasons. There is also an important requirement that the variety be maintained throughout the duration of protection. A country may apply the system to all genera or species, but there is no obligation to do so and thus the system has been extended only gradually. In addition, the UPOV Convention allows national legislation to discriminate against foreigners (including nationals of a UPOV Convention country) under the principle of reciprocity. Thus amongst the UPOV members there is still some disparity in protection. Duration of protection depends on national legislation and on the plant species to which the variety belongs, but is generally for 20–30 years. Grant of plant variety rights confers certain exclusive rights on the holder, including the exclusive right to sell the reproductive material (e.g., seed, cuttings, whole plants) of the protected variety. However the rights do not extend to consumption material (e.g., fruit, wheat seed grown for milling flour). Essentially the exclusive rights define what others may or may not do in relation to the protected varieties.

Plant breeders were for sometime dissatisfied with the protection provided by the UPOV system. This eventually resulted in a major diplomatic conference in March 1991, at which the UPOV Convention was substantially revised. The new 1991 text will provide far greater protection than is afforded at present, most notably by requiring that all member countries apply the convention to all genera and species, by extending the exclusive rights to include harvested material (e.g., fruit, wheat grown for milling into flour) and most controversially, by allowing enforcement against farm-saved seed (where a farmer produces further seed of the protected variety from the previous years crop). However, until the national governments ratify the new convention the system will continue to be based on the 1978 text. There will be considerable national opposition to the strengthening of plant variety rights and thus these changes may take years before they are implemented and may even be superseded by greater availability of patent protection in the meantime.

28.4   Patents for biotechnology

A patent is a grant of exclusive rights for a limited time in respect of a new and useful invention. The exact requirements for grant of a patent, the scope of protection it provides and its duration differs depending on national legislation. However, generally the invention must be of patentable subject matter, novel (new), non-obvious (inventive), of industrial application and sufficiently disclosed. A patent will provide a wide range of legal rights, including the right to possess, use, transfer by sale or gift and to exclude others from similar rights. Duration will be for around 20 years (although for only 17 years in the U.S.). These rights are generally restricted to the territorial jurisdiction of the country granting the patent and thus an inventor wishing to protect his/her invention in a number of countries will need to seek separate patents in each of those countries. Whilst the majority of countries provide some form of patent protection, only a few provide patent protection for biotechnology (these include: Australia, Bulgaria, Canada, Czechoslovakia, Hungary, Romania, Japan, the Soviet Union and the parties to the European Patent Convention). The reasons for this may differ, but generally it has been because biotechnology has been thought inappropriate for patent protection, either because the system was originally designed for mechanical inventions, or for technical or practical reasons, or for one or more ethical, religious or social concerns. In all the National Patent Offices where patents are granted for biotechnology there is a considerable backlog of pending applications. Even in those countries where patent protection is provided, the type and extent of that protection is different in nearly every national system.

It has largely been the U.S. which has broken new ground in providing the possibility of patent protection for 'anything under the sun that is made by man'. Patents have been granted for plants since 1930 in the U.S., under The Plant Patent Act. However, prior to 1980, the U.S. Patent Office would not grant utility patents (separate from The Plant Patent Act) living matter because it deemed products of nature not to be within the terms of the utility patent statute. That was until the landmark decision of the U.S. Supreme Court in Diamond versus Chakrabarty (from which the above quote is taken), which held that a particular genetically engineered bacterium was statutory subject matter for a utility patent. This decision has been the basis upon which patents have been granted for higher life forms. Subsequently it has been held that a utility patent may be granted for plants and a patent has been granted for an animal. Polyploid oysters, not naturally occurring, were held to be patentable subject matter and U.S. Patent No.3,736,866, was issued in respect of a 'transgenic nonhuman mammal all of whose germ cells and somatic cells contain a recombinant activated oncogene sequence introduced into the said mammal, or an ancestor of said animal, at an embryonic stage' — popularly known as the 'onco-mouse'.

Elsewhere, the treatment of applications for patents for living matter is far from certain. Whilst patents are granted in many countries for plants and micro-organisms, it has been the issue of patents for animals which has been most controversial. Whilst it is not possible to summarise succinctly the position in the rest of the world, it is possible to describe the present approach of those countries which are party to the European Patent Convention. The EPC is a regional arrangement entered into by 14 European countries for the purpose of making multiple applications for any of the member countries a great deal easier and to introduce a common system for patent protection. An application under the EPC is for a European patent, or Europatent, for short. If a Europatent is granted by the European Patent Office (EPO) it has the same effect and is subject to the same conditions, as a national patent in each of the member countries designated in the application. In other words, through a single application a bundle of national patents can be obtained.

The EPC provides that 'plant or animal varieties or essentially biological processes for the production of plants or animals' are excluded from patent protection (although the exclusion is expressly stated not to apply to micro-biological processes and products). These exclusions would appear to place unequivocal prohibition on Europatents for macrobiotechnology. However, the EPO has been taking an increasingly narrow view of these exclusions and has held that they do not exclude all plants and animals *per se*, but only claims for varieties of plants or animals and that a process is not 'essentially biological' if there has been substantial interference by man. It is also important to note that there is currently before the European Parliament of the European Community (EC) a proposal for a Council Directive for harmonisation of the legal protection provided for biotechnology in the EC. This does not propose to amend the EPC, but the present draft proposal would make even more opportunities available for patenting biotechnology and thus make the EC more attractive in terms of investment in biotechnology research.

28.5   International intellectual property treaties

There are three international intellectual property treaties which are of particular importance for the protection of biotechnology: the Paris Convention for the Protection of Industrial Property (the Paris Convention); the Budapest Treaty on the International Recognition of the Deposit of Micro-organisms for the Purposes of Patent Procedure (the Deposit Treaty) and the Patent Cooperation Treaty (PCT). The Paris Convention was originally signed in 1883 by just 11 countries, but now the majority of countries who have any form of intellectual property law are parties to it.

The keystone to the convention is the principle of national treatment: An applicant from one convention country shall have the same rights in a second convention country as a national of that second country. The convention covers patents and defines them so broadly that it permits application to any of the forms of industrial patents granted under the laws of the convention countries. The most important practical result of the convention is that it is possible to claim priority from an application made in a convention country for all subsequent convention countries within 12 months of the original filing.

The Deposit Treaty, as the full title suggests, is concerned with the deposit of examples of micro-organisms for the purposes of patent applications. Applications for patents for biotechnology often face considerable difficulties in describing the nature of the invention sufficiently. The Deposit Treaty is a vehicle for solving these problems, primarily through the setting up of a series of International Depository Authorities (IDA) and through the recognition by all member countries of a deposit in a single IDA.

The PCT simplifies the process of filing patent applications simultaneously in a number of countries. Under the PCT a single application may be filed in one of the official receiving offices, designating any number of PCT member countries, which can eventually result in a national patent being granted in each of the designated states (and/or a Europatent). A prior-art search is performed by the receiving office and a report sent to the applicant. The application and report are published and the application will then move on either to an international preliminary examination followed by national examination, or alternatively straight to the national examination stage. Unfortunately, the eventual outcome is not a 'world patent' and there is no harmonisation patent law under the PCT apart from the procedural aspects.

28.6 Pitfalls of proposed global patenting system

The World Intellectual Property Organisation (WIPO), based in Geneva, Switzerland, is one of the 16 specialised agencies of the United Nations. Today, it has 184 members and administers 24 international treaties. WIPO was created in 1967, succeeding BIRPI (Bureaux Internationaux Reunis pour la Protection de la Propriete Intellectuelle), which was established in 1893 to administer the Berne Convention. Its original objective was to protect literary and artistic work. It also dealt with the Paris Convention for the protection of intellectual property. WIPO operates on 'one-member, one-vote' basis, irrespective of the population or contribution of the member state. As a result, an attempt by developed nations for universal pharmaceutical patents was blocked by developing nations in 1960s and 1970s. This led to the U.S. and other developed countries shifting intellectual property standard setting out of WIPO to GATT and subsequently to Trade Related Aspects of Intellectual Property Rights (TRIPS) under World Trade Organisation (WTO).

While the TRIPS agreement deals with a large number of issues like copyright, geographical indications, industrial designs, integrated circuit layout design, monopolies for new plant varieties developed, trademarks, etc., its major focus still continues to be intellectual property rights through the patent system.

28.6.1 Patent protection under TRIPS

Agreement on TRIPS sets down minimum standards for intellectual property regulation. It is an international agreement administered by the WTO. It was negotiated at the end of the Uruguay Round of the General Agreement on Tariffs and Trade (GATT) in 1994. TRIPS is a comprehensive international agreement that specifies registering and enforcement procedures, dispute regulation procedures, etc., so that promotion of technological innovation and its transfer is regulated to mutual advantage of both the producer and the user of technical knowledge in a manner conducive to social and economical welfare. It balances rights and obligations. Taking into account the concerns of developing countries, a 10-year grace period was given to them, which expired in 2005. For the least developed countries, this period is up to 2019 and could be extended even beyond that. Safeguards to protect public health have been incorporated into the TRIPS agreement; however, in practice governments were reluctant to exercise such rights, given the concern about the international trade and political ramifications. The objective of the Doha Declaration was to promote access of medicines to all. This has to be viewed particularly in the light of the AIDS pandemic in Africa and high cost of AIDS drugs. In spite of the safeguards in TRIPS, many developing countries did not take advantage of the TRIPS flexibilities in terms of compulsory licensing, parallel importation, data protection, etc.

This happened mainly due to lack of legal and technical expertise in developing countries. It is in this respect that the WIPO has relevance.

WIPO works through setting up Standing Committees: it has a committee for patents, copyright and related rights; advisory committee for enforcement; intergovernmental committee on access to genetic resources; committee for traditional knowledge, folklore, etc. Patent Cooperation Treaty (PCT) is one such initiative of WIPO.

28.6.2 Patent co-operation treaty

To prevent unauthorised copying of intellectual property, the PCT was concluded in 1970. It came into effect in 1978 with 18 member states. Under the treaty, the first international application was filed in 1978. The treaty has since been amended in 1979, 1984 and 2001. PCT facilitates the obtaining of protection for innovation in any or all PCT member countries. Application under PCT avoids filing several separate national and/or regional patent applications under several regional patent treaties such as African Regional Intellectual Property Organisation, Eurasian Patent Convention, European Patent Convention, etc. PCT also facilitates processing of applications nationally. Under the treaty, a single filing by the member state is made at the Receiving Office. A search is carried out by the International Searching Authority and written opinion given regarding patentability of the invention.

This is followed by preliminary examination by the International Preliminary Examining Authority (IPEA) under the PCT. All this, however, does not lead to granting of International Patent. The main advantage of PCT is to give time to the member state, avoiding translation costs and avoiding national or regional procedures. The first phase comprises of giving the applicant time within which its patent rights are protected in member states.

On filing an application at the PCT, the International Search Authority takes anything from 9–16 months before it can submit International Search Report (ISR), together with written opinion regarding patentability. ISR is then published. The ISR not only helps the applicant in filing the patent nationally, but also in deciding whether it is worthwhile filing patents in other countries and if so, how many. Further, as many national patent offices trust ISRs, in the process, an applicant saves lot of time. ISR search and opinion given by it does not affect whether or not the invention is patentable locally. In fact PCT Article 27(5) clearly states, 'Nothing in this Treaty and the Regulations is intended to be construed as prescribing anything that would limit the freedom of each Contracting State to prescribe such substantive conditions of patentability as it desires'. The international phase ends at 30 months from the filing of application under PCT. Utility of the PCT is borne out by the fact that the millionth patent application under PCT was filed by end of 2004. But now, there is an attempt to change the objective and spirit of PCT.

28.6.3 Patent prosecution highway under PCT

There are apprehensions that ever since Dr. Claus Matthes joined as Director at Patent Cooperation Organisation, he has been propagating a global patenting system. Patent cooperation system is administered by WIPO and is also a procedural treaty as reiterated by Dr. Matthes recently in a leading Indian newspaper. He claims that his objective is not to have global patenting system or to dilute the sovereignty of developing nations.

According to Dr. Claus Matthes, the objective is to reduce work load and avoid duplication of work in member countries, thereby reducing cost. He wants to improve quality of search and examination work. It is in this context that he is proposing expanding of Patent Prosecution Highway (PPH) to enable PCT to function more effectively. There is also the proposed Substantive Patent Law Treaty (SPLT) aimed at harmonising substantive points of patent law.

To achieve the above stated objectives, Dr. Matthes wants an in-depth study to be done for improving the functioning of PCT by identifying existing problems and challenges; analysing and underlining causes of prevailing problems; finding possible solutions; and evaluating their impact. The study report is to be presented to a PCT working group in 2010.

28.6.4 Objections to PCT reforms

Dr. Matthes' proposals have raised strong objections. Dr. Y.K. Hamied, Chairman and Managing Director, Cipla Ltd., has noted that the WTO-TRIPS Agreement has given freedom to each country to decide and legislate their own standard for patentability, including provision of compulsory licensing. Section 3(d) of

TRIPS addresses the medical need of a country and prevents patentee from extending patent term through ever-greening of patent without real novelty or innovation. It prevents frivolous patents being granted. Multinational companies have challenged the provisions of Section 3(d) in the international courts, but were frustrated by landmark court judgements. India has been producing and offering affordable medicines to developing countries and the MNCs are trying to undo the same by proposing to modify the existing TRIPS provisions by proposing Universal Patent Laws through the PCT. '0.6 billion people in developed countries want to deny cheap medicines to 3 billion people in developing countries', laments Dr. Hamied. The developed countries want to continue existing monopolies at the cost of Third World and developing countries, thereby denying them access to medicines at affordable prices, he asserts.

A number of public interest organisations such as Health Equity and Society, Indian Network for People Living with HIV/AIDS, All India Drug Action Network, Diverse Women for Diversity, International Peoples Health Council (South Asia), Drug Action Forum — Karnataka, Centre for Legislative Research and Advocacy, SATHI-CEHAT, Knowledge Commons, International Treatment Preparedness Coalition-India, Delhi Science Forum, All India People's Science Network, LOCOST, Centre for Health and Social Justice, National Working Group on Patent Laws, IT for Change, Centre for Internet and Society and Centre for Trade and Development have appealed to the Indian Government to oppose any efforts at WIPO General Assembly on harmonisation of patent administration. The appeal questions the rationale of establishing a new Committee on WIPO Standards (CWS) and Committee on Global IP Infrastructure (CGI).

The appeal further requests the government to retain the provisions available under TRIPS Agreement and Doha Declaration. In this connection, it is worth mentioning that under the above provisions, 'Tamiflu' for treatment of H1N1 flu (Swine flu) has become generic in India and therefore available at reasonable prices. It is also under the present patent regime operating in India that generic anti-AIDS drugs such as 'Tenofovir', 'Darunavir' and leukaemia drug imatinib mesylate (Glivec) are available in India.

Unfortunately, 92.5% of PCT applications in 2008 were from 15 developed countries. These are the one which generate revenues and WIPO secretariat depends on them for their livelihood. According to Mr. D.G. Shah, Secretary General, Indian Pharmaceutical Alliance, Dr. Matthes' predecessors focused on issues like traditional knowledge and capacity building. Now Dr. Matthes is pushing for new priorities, which would dilute the development agenda, reducing scope of capacity building programmes, discarding programmes for SMEs, curtailing budget of WIPO academy, says Mr. Shah. He feels that the focus should be on following aspects:

1. Balanced evolution of the international normative framework for IP.

2. Provision of premier global IP services.

3. Facilitating the use of IP for sustainable development.

4. Coordination and development of global IP infrastructure.

Dr. Matthes' proposals would favour developed countries by expanding their dominance, asserts Mr. Shah. The Substantive Patent Law Treaty proposed earlier was blocked by developing countries since it sought to harmonise patent laws. Dr. Matthes now wants to go beyond the above rejected law by doing away with search and examination in the national phase. This will result in automatic patent being granted in all member states. If his objective is achieved, the patent offices in developing countries will become redundant.

28.7   Case study: The Iguana management programme

The Green Iguana of Latin America is a highly prized source of meat and eggs. Green Iguanas are arboreal herbivores which can grow up to 2 m in length and can weigh as much as 6 kg (about 82% of the lizard is edible). They need about half as much food as a chicken or rabbit to produce the same amount of meat. The species is now widely threatened because of excess hunting and habitat destruction.

Research into the reproductive behaviour of the Green Iguana was begun in 1983 and resulted in development of new management techniques for ranching. A 'genetic brood stock' of adult iguanas which are larger, faster growing and more productive has been developed. The research has largely been the work of the Pro Iguana Verde Foundation (formed by Dagmar Werner in 1985). The Foundation's programme for training and advice on Iguana ranching is called the Iguana Management Programme (IMP). The IMP is based in Costa Rica but it is intended to implement it throughout Latin America and possibly elsewhere.

The primary purpose of the IMP is to conserve living natural resources; its basic premise is that if farmers can raise iguanas as a food crop, the status of the wild species will be improved and forest clearance might be reduced. Farmers adopting iguana ranching would have to protect or re-establish areas of forest to provide food for stock. Research indicates that meat production per hectare by iguanas is approximately three times higher than by cattle. Income can be derived from selling iguanas and their products (meat, eggs, leather) and products from the forest.

The new technology and expertise which have been incorporated into an iguana ranching model are being applied for an industrial purpose (i.e., agriculture) and are of commercial value; they thus fall within the area of intellectual property law as applied to biotechnology. The biotechnological components of the ranching model are the genetic brood stock (the Fundacion has 'bioengineered' an improved stock of Green Iguanas) and the husbandry procedures (egg laying and incubation, nutrition, disease control, release and harvesting). These are forms of 'original or traditional biotechnology', as opposed to 'new biotechnology' which is largely laboratory-based and dependent upon human manipulation of genetic material.

Intellectual property rights provide the means for compensating the Fundacion for its efforts. The technologies involved in the IMP are vulnerable to piracy. Much of the work of the Fundacion is contained in the genetic make-up of the Genetic Brood Stock. Once these Iguanas are transferred or sold the Fundacion loses its direct control over the animals. In addition, the success of the Iguana ranching model is dependent on the expertise to use the technologies efficiently; this is information which took years to develop but which can be pirated very easily once a license is purchased. The Fundacion needs to be able to disseminate its innovations and expertise in the security of knowing that it cannot be resold by pirates and that there will be no reduction of the licensing potential. Only internationally recognised intellectual property law can provide these types of protection.

Because of the uncertainties of the world's intellectual laws with regard to biotechnology the availability of protection for the most important components of the IMP is questionable. At present there is widespread discrimination against the application of intellectual property rights to natural genetic materials and in favour of human-modified genetic materials. This provides no incentives for exploitation of useful genetic materials in the natural environment, even though in developing countries natural resources are obvious subjects for investment. However, one important way to limit conversion of natural resources is to ensure that fair value is paid for current uses of the existing resource base. Intellectual property rights could be a means of influencing developing countries to maintain and develop diverse resources in return for the value that these resources render to the world community.

28.8 Regulated trading in wildlife products

Regulated trading in wildlife products has the capacity of returning benefits to the users of natural habitats. It could do this if the trade were regulated in such a way as to support prices, much as is done at present with respect to agricultural commodities, where price supports provide incentives for maintaining land in its current state, as opposed to converting it to other purposes.

At present, there is no regulated trading mechanism of exactly this nature. There are, however, a number of existing international agreements which do seek to regulate trade in wildlife products. Early examples are the Western Hemisphere Convention and the 1950 Paris International Convention for Protection of Birds. These simply outlined in broad terms an obligation to control trade in wildlife products but created little structure within which these controls could be implemented. Both conventions consequently became 'sleeping treaties'. Undoubtedly the most important and effective convention which places some control on the economic exploitation of wildlife products and thereby protects biological diversity is the Convention on International Trade in Endangered Species of Wild Fauna and Flora (CITES).

28.8.1   Evolution of CITES

CITES is the most widely accepted of international treaties on the conservation of natural resources. The number of Parties has been steadily increasing from the initial signing of the convention in 1973 to a total of 113 in 1992. The convention attempts to prevent commercial trade in species of wildlife which are in danger of extinction and to control the trade in species which might become so if their trade was allowed to continue unchecked.

These data are then compiled on a computer database and in this way it is possible to determine the global levels of trade in each species. At a fine level of resolution, the trade emanating from each range state can then be compared with what is known about the wild population in that country to enable an estimation of whether it is sustainable or whether it might be detrimental to its survival. At a coarser scale, the data can show long-term trends in trade levels or trade routes, which can be used to help in understanding and therefore controlling the trade.

The convention covers not only live animals and plants but also products and derivatives of the species listed. These range from whole skins and manufactured leather products, through ivory carvings, tortoiseshell jewellery, meat, seeds and feathers to medicinal products extracted from plants such as ginseng. This causes problems for the implementation of the convention because it is necessary for enforcement of officers to determine not only what species the product is derived from but also whether the species is included in the appendices.

Abundance class: Refers to the relative abundance of different mRNA molecules in a cell at any given time.

ADA: Adenosine deaminase, deficiency results in SCIDS.

Adaptor: A synthetic single-stranded non self-complementary oligo-nucleotide used in conjunction with a linker to add cohesive ends to DNA molecules.

Adenine (A): Nitrogenous base found in DNA and RNA

Adeno-associated virus: Virus used in gene therapy delivery methods.

Adenovirus: Virus that can infect through nasal passages, used in gene therapy delivery methods.

Aetiology: Of disease; relating to the causes of the disease.

Agrobacterium tumefaciens: Bacterium that infects plants arid causes crown gall disease. Carries a plasmid used for gene manipulation in plants.

Agarose: Jellylike matrix, extracted from seaweed, used as a support in the separation of nucleic acids by gel electrophoresis.

Alkaline phosphatase: An enzyme that removes 5′ phosphate groups from the ends of DNA molecules, leaving 5′ hydroxyl groups.

Allele: One of two or more variants of a particular gene.

Allele-specific oligonucleotide: Oligonucleotide with a sequence that can be matched precisely to a particular allele by using stringent hybridisation conditions.

Alpha-peptide: Part of the β-galactosidase protein, encoded by the *lacZ′* gene fragment.

Ampicillin (Ap): A semisynthetic β-lactam antibiotic.

Aneuploidy: Variation in chromosome 'number where single chromosomes are affected, thus the chromosome complement is not an exact multiple of the haploid chromosome number.

Animal model: Usually a transgenic mouse in which a disease state has been engineered.

Antibody: An immunoglobulin that specifically recognises and binds to an antigenic determinant on an antigen.

Anticodon: The three bases on a tRNA molecule that are complementary to the codon on the mRNA.

Antigen: A molecule that is bound by an antibody. Also used to describe molecules that can induce an immune response, although these are more properly described as immunogens.

Antiparallel: The arrangement of complementary DNA strands, which run in different directions with respect to their 5'–3' polarity.

Antisense RNA: Produced from a gene sequence inserted in the opposite orientation, so that the transcript is complementary to the normal mRNA and can therefore bind to it and prevent translation.

Arabidopsis thaliana: Small plant favoured as a research organism for plant molecular biologists.

Autoradiograph: Image produced on X-ray film in response to the emission of radio active particles.

Autosome: A chromosome that is not a sex chromosome.

Auxotroph: A cell that requires nutritional supplements for growth.

Bacillus thuringiensis: Bacterium used in crop protection, and in the generation of *Bt* plants that are resistant to insect attack. The bacterium produces a toxin that affects the insect.

Bacteriophage: A bacterial virus.

Baculovirus: A particular type of virus that infects insect cells, producing large inclusions in the infected cells.

Bal 31 nuclease: An exonuclease that degrades both strands of a DNA molecule at the same time.

Bioinformatics: The emerging discipline of collating and analysing biological information, especially genome sequence information.

Biolistic: Refers to a method of introducing DNA into cells by bombarding them with microprojectiles, which carry the DNA.

Blunt ends: DNA termini without overhanging 3' or 5' ends.

Bovine somatotropin (BST): Bovine growth hormone, produced as rBST for use in dairy cattle to increase milk production.

Bt plants: Plants which carry the toxin-producing gene from *Bacillus thuringiensis* as a means to protect the plant from insect attack.

Callus: A mass of relatively unspecialised tissue used in plant tissue culture as the starting material for the propagation of plant clones.

Carboxyl (C) terminus: Carboxyl terminus, defined by the –COOH group of an amino acid or protein.

CAAT box: A sequence located approximately 75 base-pairs upstream from eukaryotic transcription start sites. This sequence is one of those that enhance binding of RNA polymerase.

Caenorhabditis elegans: A nematode worm used as a model organism in developmental and molecular studies.

Cap: A chemical modification that is added to the 5′ end of a eukaryotic mRNA molecule during post-transcriptional processing of the primary transcript.

Capsid: The protein coat of a virus.

cDNA: DNA that is made by copying mRNA using the enzyme reverse transcriptase.

cDNA library: A collection of clones prepared from the mRNA of a given cell or tissue type, representing the genetic information expressed by such cells.

Central dogma: Statement regarding the unidirectional transfer of information from DNA to RNA to protein.

CFTR gene (protein): Cystic fibrosis transmembrane conductance regulator, the gene and protein involved in defective ion transport that causes cystic fibrosis.

Chimaera: An organism (usually transgenic) composed of cells with different genotypes.

Chromosome: A DNA molecule carrying a set of genes. There may be a single chromosome, as in bacteria, or multiple chromosomes, as in eukaryotic organisms.

Chromosome jumping: Technique used to isolate non-contiguous regions of DNA by 'jumping' across gaps that may appear as a consequence of uncloned regions of DNA in a gene library.

Chromosome walking: Technique used to isolate contiguous cloned DNA fragments by using each fragment as a probe to isolate adjacent cloned regions.

Chymosin (chymase): Enzyme used in cheese production, available as recombinant product.

Cis-acting element: A DNA sequence that exerts its effect only when on the same DNA molecule as the sequence it acts on. For example, the CAAT box is a *cis*-acting element for transcription in eukaryotes.

Cistron: A sequence of bases in DNA that specifies one polypeptide.

Clone: (1) A colony of identical organisms; often used to describe a cell carrying a recombinant DNA fragment, (2) Used as a verb to describe the generation of recombinants, (3) A complex organism (e.g. sheep) generated from a totipotent cell nucleus by nuclear transfer into an enucleated ovum.

Codon: The three bases in mRNA that specify a particular amino acid during translation.

Cohesive ends: Those ends (termini) of DNA molecules that have short complementary sequences that can stick together to join two DNA molecules. Often generated by restriction enzymes.

Competent: Refers to bacterial cells that are able to take up exogenous DNA.

Competitor RT-PCR: Technique used to quantify the amount of PCR product by spiking samples with known amounts of a competitor sequence.

Complementation: Process by which genes on different DNA molecules interact. Usually a protein product is involved, as this is a diffusible molecule that can exert its effect away from the DNA itself. For example, a *lacZ+* gene on a plasmid can complement a mutant (*lacZ–*) gene on the chromosome by enabling the synthesis of β-galactosidase.

Concatemer: A DNA molecule composed of a number of individual pieces joined together via cohesive ends.

Congenital: Present at birth, usually used to describe genetically derived abnormalities.

Conjugation: Plasmid-mediated transfer of genetic material from a 'male' donor bacterium to a 'female' recipient.

Consensus sequence: A sequence that is found in most examples of a particular genetic element, and which shows a high degree of conservation. An example is the CAAT box.

Copy number: (1) The number of plasmid molecules in a bacterial cell, (2) The number of copies of a gene in the genome of an organism.

cos site: The region generated when the cohesive ends of l DNA join together.

Cosmid: A hybrid vector made up of plasmid sequences and the cohesive ends (*cos* sites) of bacteriophage lambda.

Crown gall disease: Plant disease caused by the Ti plasmid of *Agrobacterium tumefaciens*, in which a 'crown gall' of tissue is produced after infection.

Cyanogen bromide: Chemical used to cleave a fusion protein product from the N-terminal vector-encoded sequence after synthesis.

Cystic fibrosis: Disease affecting lungs and other tissues, caused by ion transport defects in the cm gene.

Cytosine (C): Nitrogenous base found in DNA and RNA.

Deletion: Change to the genetic material caused by removal of part of the sequence of bases in DNA.

Deoxynucleoside triphosphate (dNTP): Triphosphorylated (high energy) precursor required for synthesis of DNA, where N refers to one of the four bases (A,G,T or C).

Deoxyribonucleic acid (DNA): A condensation heteropolymer composed of nucleotides. DNA is the primary genetic material in all organisms apart from some RNA viruses. Usually double-stranded.

Deoxyribose: The sugar found in DNA.

Deoxyribonuclease (DNase): An nuclease enzyme that hydrolyses (degrades) single- and double-stranded DNA.

Dicotyledonous plant: Plant which develops from two cotyledons in the seed.

Dideoxynucleoside triphosphate (ddNTP): A modified form of dNTP used as a chain terminator in DNA sequencing.

Diploid: Having two sets of chromosomes.

Disarmed vector: A vector in which some characteristic (e.g., conjugation) has been disabled.

DNA chip: A DNA microarray used in the analysis of gene structure and expression. Consists of oligonucleotide sequences immobilised on a chip array.

DNA footprinting: Method of identifying regions of DNA to which regulatory proteins will bind.

DNA ligase: Enzyme used for joining DNA molecules by the formation of a phosphodiester bond between a 5′ phosphate and a 3′ OH group.

DNA polymerase: An enzyme that synthesises a copy of a DNA template.

DNA profiling: Term used to describe the various methods for analysing DNA to establish identity of an individual.

Dominant: An allele that is expressed and appears in the phenotype in heterozygous individuals.

Dot-blot: Technique in which small spots, or dots, of nucleic acid are immobilised on a nitrocellulose or nylon membrane for hybridisation.

Downstream processing: Refers to the procedures used to purify products (usually proteins) after they have been expressed in bacterial, fungal or mammalian cells.

Drosophila melanogaster: Fruit fly used as a model organism in genetic, developmental and molecular studies.

Duchenne muscular dystrophy: X-linked muscle-wasting disease caused by defects in the gene for the protein dystrophin.

Dystrophin: Large protein linking the cytoskeleton to the muscle cell membrane, defects in which cause muscular dystrophy.

Ectopic: Occurring in an unusual place or in an unusual form or manner.

Electroporation: Technique for introducing DNA into cells by giving a transient electric pulse.

ELSI: Sometimes used as shorthand to describe the ethical, legal and social implications of genetic engineering.

Embryo splitting: Technique used to clone organisms by separating cells in the early embryo, which then go on to direct development and produce identical copies of the organism.

End labelling: Adding a radioactive molecule onto the end(s) of a polynucleotide.

Endonuclease: An enzyme that cuts within a nucleic acid molecule, as opposed to an exonucleas, which digests DNA from one or both ends.

Enhancer: A sequence that enhances transcription from the promoter of a eukaryotic gene. May be several thousand base-pairs away from the promoter.

Enzyme: A protein that catalyses a specific reaction.

Enzyme replacement therapy: Therapeutic procedure in which a defective enzyme function is restored by replacing the enzyme itself.

Epigenesis: Theory of development that regards the process as an iterative series of steps, in which various signals and control events interact to regulate development.

Escherichia coli: The most commonly used bacterium in molecular biology.

Ethidium bromide: A molecule that binds to DNA and fluoresces when viewed under ultraviolet light. Used as a stain for DNA.

Eukaryotic: The property of having a membrane-bound nucleus.

Exon: Region of a eukaryotic gene that is expressed via mRNA.

Exonuclease: An enzyme that digests a nucleic acid molecule from one or both ends.

Explant: A piece of living tissue taken from its normal situation to a culture medium

Expressivity: The degree to which a particular genotype generates its effect in the phenotype.

Extrachromosomal element: A DNA molecule that is not part of the host cell chromosome.

Ex vivo: Outside the body. Usually used to describe gene therapy procedure in which the manipulations are performed outside the body, and the altered cells returned after processing.

Flavr Savr (sic): Transgenic tomato in which polygalacturonase synthesis is restricted using antisense technology.

Foldback DNA: Class of DNA which has palindromic or inverted repeat regions that reanneal rapidly when duplex DNA is denatured.

Fusion protein: A hybrid recombinant protein that contains vector-encoded amino acid residues at the N terminus.

β-Galactosidase: An enzyme encoded by the *lacZ* gene. Splits lactose into glucose and galactose.

Gamete: Refers to the haploid male (sperm) and female (egg) cells that fuse to produce the diploid zygote (qv) during sexual reproduction.

Gel electrophoresis: Technique for separating nucleic acid molecules on the basis of their movement through a gel matrix under the influence of an electric field.

Gel retardation: Method of determining protein-binding sites on DNA fragments on the basis of their reduced mobility, relative to unbound DNA, in gel electrophoresis experiments.

Gene: The unit of inheritance, located on a chromosome. In molecular terms, usually taken to mean a region of DNA that encodes one function. Broadly, therefore, one gene encodes one protein.

Gene cloning: The isolation of individual genes by generating recombinant DNA molecules, which are then propagated in a host cell which produces a clone that contains a single fragment of the target DNA.

Gene protection technology: Range of techniques used to ensure that particular commercially derived recombinant constructs cannot be used without some sort of control or process, usually supplied by the company marketing the recombinant. Also known as genetic use restriction technology and genetic trait control technology.

Gene therapy: The use of cloned genes in the treatment of genetically-derived malfunctions. May be delivered *in vivo* or *ex vivo*. May be offered as gene addition or gene replacement versions.

Genetic code: The triplet codons that determine the types of amino acid that are inserted into a polypeptide during translation. There are 61 codons for 20 amino acids (plus three stop codons), and the code is therefore referred to as degenerate.

Genetic fingerprinting: A method which uses radioactive probes to identify bands derived from hypervariable regions of DNA. The band pattern is unique for an individual, and can be used to establish identity or family relationships.

Genetic mapping: Low-resolution method to assign gene locations (loci) to their position on the chromosome.

Genetic marker: A phenotypic characteristic that can be ascribed to a particular gene.

Genetic trait control technology: Version of gene protection technology, sometimes called 'traitor technology'.

Genetically modified organism (GMO): An organism in which a genetic change has been engineered. Usually used to describe transgenic plants and animals.

Genome: Used to describe the complete genetic complement of a virus, cell or organism.

Genomics: The study of genomes, particularly genome sequencing.

Genomic library: A collection of clones which together represent the entire genome of an organism.

Genotype: The genetic constitution of an organism.

Germ line: Gamete producing (reproductive) cells that give rise to eggs and sperm.

Guanine (G): Nitrogenous base found in DNA and RNA.

Haploid: Having one set of chromosomes.

Heterologous: Refers to gene sequences that are not identical, but show variable degrees of similarity.

Heteropolymer: A polymer composed of different types of monomer. Most protein and nucleic acid molecules are heteropolymers.

Heterozygous: Refers to a diploid organism (cell or nucleus) which has two different alleles at a particular locus.

Homologous: (1) Refers to paired chromosomes in diploid organisms, (2) Used to strictly describe DNA sequences that are identical; however, the percentage homology between related sequences is sometimes quoted.

Homopolymer: A polymer composed of only one type of monomer, such as polyphenylalanine (protein) or polyadenine (nucleic acid).

Homozygous: Refers to a diploid organism (cell or nucleus) which has identical alleles at a particular locus.

Host: A cell used to propagate recombinant DNA molecules.

Hybrid-arrest translation: Techniques used to identify the protein product of a cloned gene, in which translation of its mRNA is prevented by the formation of a DNA mRNA hybrid.

Hybrid-release translation: Technique in which a particular mRNA is selected by hybridisation with its homologous cloned DNA sequence, and is then translated to give a protein product that can be identified.

Hybridisation: The joining together of artificially separated nucleic acid molecules via hydrogen bonding between complementary bases.

Hyperchromic effect: Change in absorbance of nucleic acids, depending on the relative amounts of single-stranded and double-stranded forms. Used as a measurement in denaturation/renaturation studies.

Hypervariable region (HVR): A region in a genome that is composed of a variable number of repeated sequences and is diagnostic for the individual.

Ice-minus bacteria: Bacteria engineered to disrupt the normal ice-forming process, used to protect plants from frost damage.

Insertion vector: A bacteriophage vector that has a single cloning site into which DNA is inserted.

Insulin-like growth factor (IGF-1): Polypeptide hormone, synthesis of which is stimulated by growth hormone. Implicated in some concerns about the safety of using recombinant bovine growth hormone in cattle to increase milk yields.

Intervening sequence: Region in a eukaryotic gene that is not expressed via the processed mRNA.

Inverted repeat: A short sequence of DNA that is repeated, usually at the ends of a longer sequence, in a reverse orientation.

In vitro: Literally 'in glass', meaning in the test-tube, rather than in the cell or organism.

In vivo: Literally 'in life', meaning the natural situation, within a cell or organism.

IPTG: iso-Propyl-thiogalactoside, a gratuitous inducer which derepresses transcription of the lac operon.

Kilobase (kb): 103 bases or base-pairs, used as a unit for measuring or specifying the length of DNA or RNA molecules.

Klenow fragment: A fragment of DNA polymerase I that lacks the 5′–3′ exonuclease activity.

Knockin mouse: A transgenic mouse in which a gene function has been added or 'knocked in'. Used primarily to generate animal models for the study of human disease.

Knockout mouse: A transgenic mouse in which a gene function has beep disrupted or 'knocked out'. Used primarily to generate animal models for the study of human disease, e.g., cystic fibrosis.

Linkage mapping: Genetic mapping technique used to establish the degree of linkage between genes.

Linker: A synthetic self-complementary oligonucleotide that contains a restriction enzyme recognition site. Used to add cohesive ends to DNA molecules that have blunt ends.

Lipase: Enzyme that hydrolyses fats (lipids).

Liposome (lipoplex): Lipid-based method for delivering gene therapy.

Locus: The site at which a gene is located on a chromosome.

Lysogenic: Refers to bacteriophage infection that does not cause lysis of the host cell.

Lytic: Refers to bacteriophage infection that causes lysis of the host cell.

Maternal inheritance: Pattern of inheritance from female cytoplasm. Mitochondrial genes are inherited in this way, as the mitochondria are inherited with the ovum.

Mega (M): SI prefix, 10^6.

Messenger RNA (mRNA): The ribonucleic acid molecule transcribed from DNA that carries the codons specifying the sequence of amino acids in a protein.

Micro (μ): SI prefix, 10^{-6}.

Microinjection: Introduction of DNA into the nucleus or cytoplasm of a cell by insertion of a microcapillary and direct injection.

Microsatellite DNA: Type of sequence repeated many times in the genome. Based on dinucleotide repeats, microsatellites are highly variable and can be used in mapping and profiling studies.

Milli (m): SI prefix, 10^{-3}.

Minisatellite DNA: Type of sequence based on variable number tandem repeats (VNTRs). Used in genetic mapping and profiling studies.

Molecular cloning: Alternative term for gene cloning.

Molecular ecology: Use of molecular biology and recombinant DNA techniques in studying ecological topics.

Molecular paleontology: Use of molecular techniques to investigate the past, as in DNA profiling from mummified or fossilised samples.

Monocistronic: Refers to a RNA molecule encoding one function.

Monogenic: Trait caused by a single gene.

Monocotyledonous plant: Plant which develops from a single cotyledon in the seed.

Monomer: The unit that makes up a polymer. Nucleotides and amino acids are the monomers for nucleic acids and proteins, respectively.

Monosomic: Diploid cells in which one of a homologous pair of chromosomes has been lost.

Monozygotic: Refers to identical twins, generated from the splitting of a single embryo at an early stage.

Mosaic: An embryo or organism in which not all the cells carry identical genomes.

mRNA: Messenger RNA transcribed from DNA in the nucleus, exported into the cytoplasm and translated into protein.

Multifactorial: Caused by many factors, e.g. genetic trait in which many genes and environmental influences may be involved.

Multi-locus probe: DNA probe used to identify several bands in a DNA fingerprint or profile. Generates the 'bar code' pattern in a genetic fingerprint.

Multiple cloning site (MCS): A short region of DNA in a vector that has recognition sites for several restriction enzymes.

Multipotent: Cell which can give rise to a range of differentiated cells.

Mutagenesis: The process of inducing mutations in DNA.

Mutant: An organism (or gene) carrying a genetic mutation.

Mutation: An alteration to the sequence of bases in DNA. May be caused by insertion, deletion or modification of bases.

Nano (n): SI prefix, 10^{-9}.

Native protein: A recombinant protein that is synthesised from its own N terminus, rather than from an N terminus supplied by the cloning vector.

Nested fragments: A series of nucleic acid fragments that differ from each other (in terms of length) by one or only a few nucleotides. Nick translation Method for labelling DNA with radioactive dNTPs.

Northern blotting: Transfer of RNA molecules onto membranes for the detection of specific sequences by hybridisation.

N terminus: Amino terminus, defined by the $-NH_2$ group of an amino acid or protein.

Nuclear transfer: Method for cloning organisms in which a donor nucleus is taken from a somatic cell and transferred to the recipient ovum.

Nuclease: An enzyme that hydrolyses phosphodiester bonds.

Nucleoside: A nitrogenous base bound to a sugar.

Nucleotide: A nucleoside bound to a phosphate group.

Nucleoid: Region of a bacterial cell in which the genetic material is located.

Nucleus: Membrane-bound region in a eukaryotic cell that contains the genetic material.

Oligo: Prefix meaning few, as in oligonucleotide or oligopeptide.

Oligo(dT)-cellulose: Short sequence of deoxythymidine residues linked to a cellulose matrix, used in the purification of eukaryotic mRNA.

Oligomer: General term for a short sequence of monomers.

Oligonucleotide: A short sequence of nucleotides.

Oligonucleotide-directed mutagenesis: Process by which a defined alteration is made to DNA using a synthetic oligonucleotide.

Oncogene: Gene involved in the formation of cancerous tissue.

Oncomouse: Transgenic mouse engineered to be susceptible to cancer.

Oocyte: Stage in development of the female gamete or ovum (egg). Often the terms oocyte and ovum are used interchangeably.

Operator: Region of an operon, close to the promoter, to which a repressor protein binds.

Operon: A cluster of bacterial genes under the control of a single regulatory region.

Organelle: Any discrete structure in an individual cell of a multicellular organism that is adapted and/or specialised for the performance of one or more vital functions.

Organismal cloning: The production of an identical copy of an individual organism by techniques such as embryo splitting or nuclear transfer. Used to distinguish the process from molecular cloning.

Ovum: The mature female gamete or egg cell, derived from the oocyte. Often the terms ovum and oocyte are used interchangeably.

Palindrome: A DNA sequence that reads the same on both strands when read in the same (e.g., 5′–3′) direction. Examples include many restriction enzyme recognition sites.

Pedigree analysis: Determination of the transmission characteristics of a particular gene by examination of family histories.

Penetrance: The proportion of individuals with a particular genotype that show the genotypic characteristic in the phenotype.

Phagemid: A vector containing plasmid and phage sequences.

Pharm animal: Transgenic animal used for the production of pharmaceuticals.

Phenotype: The observable characteristics of an organism, determined both by its genotype and its environment.

Phosphodiester bond: A bond formed between the 5′ phosphate and the 3′ hydroxyl groups of two nucleotides.

Physical mapping: Mapping genes with reference to their physical location on the chromosome. Generates the next level of detail compared to genetic mapping.

Physical marker: A sequence-based tag that labels a region of the genome. There are several such tags that can be used in mapping studies.

Pico (p): SI prefix, 10^{-12}.

Plaque: A cleared area on a bacterial lawn caused by infection by a lytic bacteriophage.

Plasmid: A circular extrachromosomal element found naturally in bacteria and some other organisms. Engineered plasmids are used extensively as vectors for cloning.

Ploidy number: Refers to the number of sets of chromosomes, e.g., haploid, diploid, triploid, etc.

Pluripotent: Cell which can give rise to a range of differentiated cells.

Polyacrylamide: A cross-linked matrix for gel electrophoresis of small fragments of nucleic acids, primarily used for electrophoresis of DNA. Also used for electrophoresis of proteins.

Polyadenylic acid: A string of adenine residues. Poly(A) tails are found at the 3′ ends of most eukaryotic mRNA molecules.

Polycistronic: Refers to an RNA molecule encoding more that one function. Many bacterial operons are expressed via polycistronic mRNAs.

Polygalacturonase: Enzyme involved in pectin degradation. Target for antisense control in the Flavr Savr tomato.

Polygenic trait: A trait determined by the interaction of more than one gene, e.g., eye colour in humans.

Polyhedra: Capsid structures in baculoviruses, composed of the protein polyhedrin.

Polymer: A long sequence of monomers.

Polymerase: An enzyme that synthesises a copy of a nucleic acid.

Polymerase chain reaction (PCR): A method for the selective amplification of DNA sequences. Several variants exist for different applications.

Polymorphism: Refers to the occurrence of many allelic variants of a particular gene or DNA sequence motif. Can be used to identify individuals by genetic mapping and DNA profiling techniques.

Polynucleotide: A polymer made up of nucleotide monomers.

Polynucleotide kinase (PNK): An enzyme that catalyses the transfer of a phosphate group onto a 5′-hydroxyl group.

Polypeptide: A chain of amino acid residues.

Polystuffer: An expendable stuffer fragment in a vector that is composed of many repeated sequences.

Positional cloning: Cloning genes for which little information is available apart from their location on the chromosome.

Preformationism: Refers to the idea that all development is pre-coded in the zygote, and that development is simply the unfolding of this information. Now considered too simplistic.

Pribnow box: Sequence found in prokaryotic promoters that is required for transcription initiation. The consensus sequence is TATAAT.

Primary transcript: The initial, and often very large, product of transcription of a eukaryotic gene. Subjected to processing to produce the mature mRNA molecule.

Primer extension: Synthesis of a copy of a nucleic acid from a primer. Used in labelling DNA and in determining the start site of transcription.

Probe: A labelled molecule used in hybridisation procedures.

Proinsulin: Precursor of insulin that includes an extra polypeptide sequence that is cleaved to generate the active insulin molecule.

Prokaryotic: The property of lacking a membrane-bound nucleus, e.g. bacteria such as *E. coli.*

Promoter: DNA sequence(s) lying upstream from a gene, to which RNA polymerase binds.

Pronucleus: One of the nuclei in a fertilised egg prior to fusion of the gametes.

Prophage: A bacteriophage maintained in the lysogenic state in a cell.

Protease: Enzyme that hydrolyses polypeptides.

Protein: A condensation' (dehydration) heteropolymer composed of amino acid residues linked together by peptide bonds to give a polypeptide.

Proteome: Refers to the population of proteins produced by a cell.

Protoplast: A cell from which the cell wall has been removed.

Prototroph: A cell that can grow in an unsupplemented growth medium.

Purine: A double-ring nitrogenous base such as adenine and guanine.

Pyrimidine: A single-ring nitrogenous base such as cytosine, thymine and uracil.

Random amplified polymorphic DNA: PCR-based method of DNA profiling that involved amplificatio of sequences using random primers. Generates a type of genetic fingerprint that can be used to identify individuals.

Reading frame: The pattern of triplet codon sequences in a gene. There are three reading frames, depending on which nucleotide is the start .point. Insertion and deletion mutations can disrupt the reading frame and have serious consequences, as often the entire coding sequence becomes nonsense after the point of mutation.

Recessive: An allele where the expression is masked in the phenotype in heterozygous individuals.

Recombinant DNA: A DNA molecule made up of sequences that are not normally joined together.

Recombination frequency mapping: Method of genetic mapping that uses the number of crossover events that occur during meiosis to estimate the distance between genes.

Regulatory gene: A gene that exerts its effect by controlling the expression of another gene.

Renaturation kinetics: method of analysing the complexity of genomes by studying the patterns obtained when DNA is denatured and allowed to renature.

Repetitive sequence: A sequence that is repeated a number of times in the genome.

Replacement vector: A bacteriophage vector in which the cloning sites are arranged in pairs, so that the section of the genorne between these sites can be replaced with insert DNA.

Replication: Copying the genetic material during the cell cycle. Also refers to the synthesis of new phage DNA during phage multiplication.

Replicon: A piece of DNA carrying an origin of replication.

Reporter gene: A gene used to disclose the function of potential regulatory DNA sequences upstream of the reporter gene

Restriction enzyme: An endonuclease that cuts DNA at sites defined by its recognition sequence.

Restriction fragment: A piece of DNA produced by digestion with a restriction enzyme.

Restriction fragment length polymorphism (RFLP): A variation in the locations of restriction sites bounding a particular region of DNA, such that the fragment defined by the restriction sites may be of different lengths in different individuals.

Restriction mapping: Technique used to determine the location of restriction sites in a DNA molecule.

Retrovirus: A virus that has an RNA genome that is copied into DNA during the infection.

Reverse transcriptase: An RNA-dependent DNA polymerase found in retroviruses, used *in vitro* for the synthesis of cDNA.

Ribonuclease (RNase): An enzyme that hydrolyses RNA.

Ribonucleic acid (RNA): A condensation heteropolymer composed of ribonucleotides.

Ribosomal RNA (rRNA): RNA that is part of the structure of ribosomes.

Ribosome: The 'jig' that is the site of protein synthesis. Composed of rRNA and proteins.

Ribosome-binding site: A region on an mRNA molecule that is involved in the binding of ribosomes during translation.

RNA processing: The formation of functional RNA from a primary transcript. In mRNA production this involves removal of introns, addition of a 5′ cap and polyadenylation.

S1 mapping: Technique for determining the start point of transcription.

S1 nuclease: An enzyme that hydrolyses (degrades) single-stranded DNA.

Saccharomyces cerevisiae: Unicellular yeast (baker's yeast) that is extensively used as a model microbial eukaryote in molecular studies. Also used in the biotechnology industry for a range of applications, as well as in brewing and bread-making.

Screening: Identification of a clone in a genomic or cDNA library (qv) by using a method that discriminates between different clones.

SCIDS: Severe combined immunodeficiency syndrome, a condition that results from a defective enzyme (adenosine deaminase).

Scintillation counter: A machine for determining the amount of radioactivity in a sample.

Selectable marker: Gene coding for resistance to an antibiotic or herbicide which can be used to select for transformed tissue or plants.

Selection: Exploitation of the genetics of a recombinant organism to enable desirable, recombinant genomes to be selected over non-recombinants during growth.

Sequence tagged site: Refers to a DNA sequence that is unique in the genome and which can be used in mapping studies. Usually identified by PCR amplification.

Sex chromosome: The non-autosomal X and Y chromosomes in humans that determine the sex of the individual. Males are XY, females XX.

Sex-linked: Refers to pattern of inheritance where the allele is located on a sex chromosome.

Silencing: The process whereby an organism shuts down the expression of a gene.

Single nucleotide polymorphism: Polymorphic pattern at a single base, essentially the smallest polymorphic unit that can be identified.

Somatic cell: Body cell, as opposed to germ-line cell.

Southern blotting: Method for transferring DNA fragments onto a membrane for detection of specific sequences by hybridisation.

Specific activity: The amount of radioactivity per unit material, e.g., a labelled probe might have a specific activity of 10^6 counts/minute per microgram. Also used to quantify the activity of an enzyme.

Sperm: The mature male gamete.

Structural gene: A gene that encodes a protein product.

Stuffer fragment: The section in a replacement vector that is removed and replaced with insert DNA.

Tandem repeat: A repeat composed of an array of sequences repeated contiguously in the same orientation.

Taq polymerase: Thermostable DNA polymerase from the thermophilic bacterium *Thermus aquaticus*. Used in the polymerase chain reaction.

TATA box: Sequence found in eukaryotic promoters. Also known as the Hogness box, it is similar to the Pribnow box found in prokaryotes, and has the consensus sequence TATAAAT.

T-DNA: Region of Ti plasmid of *Agrobacterium tumefaciens* that can be used to deliver recombinant DNA into the plant cell genome.

Terminal transferase: An enzyme that adds nucleotide residues to the 3′ terminus of an oligo- or polynucleotide,

Temperate: Refers to bacteriophages that can undergo lysogenic infection of the host cell.

Thermal cycler: Heating/cooling system for PCR applications. Enables denaturation, primer binding and extension cycles to be programmed and automated.

Thermus aquaticus: Thermophilic bacterium from which *Taq* polymerase is purified. Other bacteria from this genus include *Thermus flavus* and *Thermus thermophilus*.

Thymine (T): Nitrogenous base found in DNA only.

Ti-plasmid: Plasmid of *Agrobacterium tumefaciens* that causes crown gall disease.

Tissue plasminogen activator (TPA): A protease that occurs naturally, and functions in breaking down blood clots. Acts on an inactive precursor (plasminogen), which is converted to the active form (plasmin). This attacks the clot by breaking up fibrin, the protein involved in clot formation.

Totipotent: A cell that can give rise to all cell types in an organism. Totipotency has been demonstrated by cloning carrots from somatic cells, and by nuclear transfer experiments in animals.

Transcription (Tc): The synthesis of RNA from a DNA template.

Transcriptional unit: The DNA sequence that encodes the RNA molecule, i.e. from the transcription start site to the stop site.

Transcriptome: The population of RNA molecules (usually mRNAs) that is expressed by a particular cell type.

Transfection: Introduction of purified phage or virus DNA into cells.

Transfer RNA (tRNA): A small RNA (~75–85 bases) that carries the anticodon and the amino acid residue required for protein synthesis.

Transformant: A cell that has been transformed by exogenous DNA.

Transformation: The process of introducing DNA (usually plasmid DNA) into cells. Also used to describe the change in growth characteristics when a cell becomes cancerous.

Transgene: The target gene involved in the generation of a transgenic organism.

Transgenic: An organism that carries DNA sequences that it would not normally have in its genome.

Translation (TI): The synthesis of protein from an mRNA template.

Trisomy: Aneuploid condition where an extra chromosome is present. Common example is the trisomy-21 condition that causes down syndrome.

Uracil (U): Nitrogenous base found in RNA only.

Variable number tandem repeat (VNTR): Repetitive DNA composed of a number of copies of a short sequence, involved in the generation of poly-morphic loci that are useful in genetic fingerprinting. Also known as hyper-variable regions.

Vector: A DNA molecule that is capable of replication in a host organism, and can act as a carrier molecule for the construction of recombinant DNA.

Virulent: Refers to bacteriophage that cause lysis of the host cell.

Virus: An infectious agent that cannot replicate without a host cell.

Western blotting: Transfer of electrophoretically separated proteins onto a membrane for probing with antibody.

Xenotransplantation: The use of tissues or organs from a non-human source for transplantation.

X-gal: 5-Bromo-4-chloro-3-indolyl-β-D-galactopyranoside: A chromo-genic substrate, for β-galactosidase; on cleavage it yields a blue-coloured product.

X-linked: Pattern of inheritance where the allele is located on the X-chromosome. In humans, this can result in males expressing recessive characters that would normally be masked in an autosomal heterozygote.

YAC: Yeast artificial chromosome, a vector for cloning very large pieces of DNA in yeast.

Zygote: Single-celled product of the fusion of a male and a female gamete. Develops into an embryo by successive mitotic divisions.

References

Allen, J.F., *Principles of Gene Manipulation*, Butterworths, London.

Benaim Pinto, C., *Microbiology and Microbial Biomass*, Prentice-Hall, London.

Bernard R. Glick and Jack J. Pasternak, *Molecular Biotechnology*, ASM Press, Washington, D.C.

Carolyn A. Dehlinger, *Molecular Biotechnology,* Jones and Bartlett Learning LLC, Burlington, MA.

Chang, J.C., *Handbook of Biochemical Engineering and Biotechnology*, John Wiley & Sons, New York.

Coolingwood, R.W., *Fermentation Technology and its Industrial Applications*, John Wiley & Sons, New York.

Desmond, S.T., *Genetic Engineering*, Cambridge University Press, New Delhi.

Downe, S.A., *Industrial Microbiology and Biotechnology*, John Wiley & Sons, New York.

Dugan, P., *Molecular Biology*, Plenum Publishing Corporation, London.

Goldman, M., *Biochemical Engineering*, Gordon and Breach, Science Publishers, New York.

Gould, G.W., *Textbook of Microbiology*, D. Van Nostrand, New York.

Harding, G., *Fundamentals of Biochemistry*, Prentice-Hall, London.

Hidy, M., *Biotechnology as an Intellectual Property*, Prentice-Hall, London.

Jackwerth, F., *Comprehensive Biotechnology*, Harcourt Brace Jovanovich, New York.

Jarvis, A., *Patents and Legal Protection for Biotechnology*, John Wiley & Sons, New York.

Jencks, W.P., *Expression of Foreign Proteins in Micro-organisms*, John Wiley & Sons, New York.

Keith Wilson and John Walker, *Principles and Techniques of Biochemistry and Molecular Biology,* Cambridge University Press, Cambridge, New York,

Kim, C.K., *Enzyme Biotechnology*, Marcel Dekker, New York.

Lewis, R., *Antibiotics-cloning of Biosynthetic Pathways*, Chilton Book Co., USA.

Mason, T., *Industrial Aspects of Biochemistry and Microbiology*, McGraw-Hill, New York.

McCaull, J. and Crossland, J., *Molecular Biology and Biotechnology*, Harcourt Brace Jovanovich, New York.

Mitchell, R., *Industrial Aspects of Biochemistry and Microbiology,* McGraw-Hill, New York.

Nuan, E., *Recent Advances in the Polymerase Chain Reactions*, Van Nostrand Reinhold, New York.

Odum, K., *Antibody Engineering and Perspective in Therapy*, W.B. Saunders and Co., New York.

Phillips, D.J.H., *Microbial Degradation of Halogenated Compounds*, Applied Science Publishers, London.

Primrose, S. B., *Molecular Biotechnology,* Blackwell Scientific Publications, London, UK.

Reid, G.K., *Microbial Transformation of Pesticides*, Reinhold Publishing Corporation, New York.

Robert, H.T., *Handbook of Biochemistry*, Butterworths, London.

Sax, R.A., *Microbial Insecticides*, Pergamon Press, Oxford, London.

Vollenweider, R.A., *Industrial Biotechnology*, Blackwell Scientific Publications, New York.

Wyatt, G.M., *Fermentation and Enzyme Technology*, Reston Publishing Co., Reston, Virginia.

Index